THE FLORA OF SOMERSET

CAPTAIN R. G. B. ROE, R.N.

PRINTED BY
BIGWOOD & STAPLE, BRIDGWATER
FOR SOMERSET ARCHAEOLOGICAL AND NATURAL
HISTORY SOCIETY
TAUNTON CASTLE
TAUNTON

1981

ISBN O 902152 12 2

Contents

LIST OF PLATES

PREFACE

It is over thirty years since an appointment to the Admiralty first brought me to live in Somerset. My wife and I had always been interested in wild flowers but we saw so many new and attractive ones on the downs and in the woods and lanes around Bath that "plant-hunting" soon became our favourite recreation and led to more serious studies. We felt the need for an up-to-date Local Flora and I hope the present book will help others to enjoy botanizing in Somerset as much as we have done.

Writing a Local Flora is a daunting task and would be impracticable without the help of a great many people; it is a great pleasure to express my thanks to all those who have given their help so freely. Encouragement and advice were always forthcoming from the Officers and from fellow members of the Botanical Society of the British Isles, many of whom took part in the County Survey on which this book is based. Previous records came from those published by past workers in the county in the books and periodicals quoted in the Bibliography and from the card indexes kept by Professor A. J. Willis, Botanical Recorder for the Bristol Naturalists' Society, and A. D. Hallam, my predecessor as Botanical Recorder for the Somerset Archaeological and Natural History Society; both Recorders were most helpful in providing information from their own experience in the county. Dr F. H. Perring and the staff of the Biological Records Centre, Monks Wood, answered numerous queries, as did the Curators of many Herbaria, including those at Kew and the British Museum (Natural History). A full list of those who contributed records for the County Survey is given later in the book and thanks are due to them all. Particular mention must be made of R. M. Burton who carried out a detailed survey of a large area round Weston-super-Mare between 1961 and 1963 and who very generously placed the full results at my disposal. Help with identifying critical species was given by the B.S.B.I. Referees and the staff of the British Museum (Natural History) and I am most grateful to them, particularly to P. D. Sell and C. E. A. Andrews (*Hieracium*), A. Newton (*Rubus*), Dr A. J. Richards (*Taraxacum*), the late Dr C. E. Hubbard (Grasses) and E. J. Clement (Aliens).

The photographs are by P. Wakely and the maps (based on the Ordnance Survey with the permission of The Controller of Her Majesty's Stationery Office, Crown Copyright Reserved) were very kindly drawn by Mrs Elizabeth McDonnell.

I am grateful, too, to the Somerset Archaeological and Natural History Society for publishing the book, and particularly to Dr R. W. Dunning and J. V. Carrington who have undertaken most of the work involved.

Of all those who helped none deserve more thanks than the small band of recorders who volunteered year after year to fill in the gaps and to search tetrads of little botanical promise often far from their homes. These were Mrs Joan Appleyard, H. W. Boon, J. V. Carrington, R. S. Cropper, C. H. Cummins, Dr A. F. Devonshire, Miss Caroline Giddens and J. G. Keylock. The hardest working of them all was my old friend Charles Howe who in fourteen years sent in over 100,000 records from 422 tetrads, almost half the county. He read the first half of this book in manuscript and approved of it; it is very sad that he did not live to see the whole book, which owes so much to him, in print. Finally I must gratefully acknowledge my wife's unfailing support in the field and her encouragement and forbearance during the preparation of the text.

List of Contributors to the Flora

In addition to those whose names are given in full in the text, the following have contributed records, either by taking part in the County Survey or by sending or publishing details of particular finds. Those who have made complete lists for one or more tetrads are marked with an asterisk. The names of earlier workers whose records are frequently cited are included, with the abbreviations used to refer to them.

Adam, P.
Andrews, Miss C. M. CMA
* Appleyard, Mrs J. JA
Bevan, J.
Biggar, Miss E. I.
Bingham, Miss C.
* Bingham, Miss J.
* Blomfield, Miss R. RB
* Bond, Dr T. E. T.
* Boon, H. W. HWB
* Bowcott, Mrs J. M.
Bowen, Dr H. J. M.
* Brassington, I. W.
Brenan, J. P. M. JPMB
Bromwich, Miss A.
* Brown, Dr D. H.
Bull, K. E.
* Burton, R. M. RMB
Cairns, T. TC
* Campbell, Dr N. G. D. NGC
* Carrington, J. V. JVC
Chaffey, N. J.
Chicken, E. EC
Clark, H.
Clement, E. J. EJC
* Cole, Mrs A.
Cole, M. E.
* Cole, Mrs O. E. OEC
Cook, N.

LIST OF CONTRIBUTORS

* Cook, R. J.
Coombe, Dr D. E. DEC
* Cornell, C. J. CJC
Couzens, P. L.
Cowley, Miss B.
* Cropper, R. S. RSC
* Cummins, C. H. CHC
Cursham, Cdr R.
Curtis, Mrs E.
Dancy, J.
David, R. W.
* Devonshire, Dr A. F. AFD
Dobbie, Dr J. L.
Drazin, Dr P. G.
Duff, A. AD
Evans, I. W. (1891-1969) IWE
Evans, T. G. TGE
* Everson, Mrs B.
Fishter, M.
* Franey, A.
Freeston, T. TF
* Gibbons, R. B. RBG
* Giddens, Miss C. J. CJG
Graham, Miss V. VG
* Gravestock, Miss I. F. IFG
Green, D. E. DG
Grenfell, A. L ALG
* Groome, L.
* Hacking, J. F. & Mrs Z. A.
Hadden, N. G. (d. ca. 1950) NGH
Hallam, A. D. ADH
Hallam, Mrs O. M. OMH
Hallam, A. D. and Mrs O. M. AD & OMH
Hamlin, E. J. (1878-1966) EJH
* Hare, Dr C. L. CLH
* Harley, Dr R. M. RMH
* Harris, E. B. EBH
* Hastings, Miss S. SH
Hawkins, Mrs G. F. C.
* Hickmott, P. A. & Mrs M. PA & MH
* Hill, Miss P. M. PMH

LIST OF CONTRIBUTORS

Holland, Mrs S. C. SCH
* Holt, O. D.
Hope-Simpson, Dr J. F.
* House, F. FH
* Houseman, F.
* Howe, C. A. (1897-1979) CAH
Hubbard, Dr C. E. (1900-1980) CEH
* Hudson, Dr M.
* Hunt, P. F. PFH
Huxter, T. J.
Jardine, N.
* Jeffery, R.
* Jessup, D. V. & Mrs E. M.
* Keylock, J. G. JGK
* Knight, Dr J. T. H. JTK
Kurzen, Miss S.
* Lane, Mrs L. B.
Lawson, P. G.
Lee, Mrs S. M.
* Lenton, Miss E. J. EJL
Lepper, J. D.
Lousley, J. E. (1907-76) JEL
Lovatt, C. M. CML
* McFarlane, M. G. MGMcF
Mack, Mrs B. M.
Macpherson, Dr P.
Makins, F. K. (d. 1956) FKM
* Mallet, Miss P. M. E. M.
Margetts, L. J. LJM
Marshall, A. D.
Marshall, Rev E. S. (1858-1919) ESM
Mason, Dr J. L.
* Messenger, K. G. KGM
Miller, W. D. (1868-1933) WDM
Morley, J. V. JVM
Moss, C. E. (1872-1931) CEM
Mullins, D. E.
Murray, Rev R. P. (1842-1908) RPM
* Nethercott, P. J. M. PJMN
Newton, A. AN
Norman, Mrs G.

LIST OF CONTRIBUTORS

Overend, Miss E. D. EDO
Palmer, J. R.
* Parrott, Mrs C.
Paskin, E.
Payne, R. M. RMP
Pearce, Mrs C.
Pearman, D. A.
* Peregrine, Dr D. H.
* Perrett, Mrs D. H. DHP
Pilkington, Miss F. M.
* Poore, N. R.
Preston, C. D.
* Pyke, Mrs M. H. G. MHP
Quirk, Miss M.
Ralph, C. D.
Randall, R. D. RDR
Rawlins, Miss E. (d. 1956) ER
Ricketts, Mrs V. I. VIR
Robinson, Miss J.
* Roe, Capt R. G. B. RGR
Roe, Mrs I. G. IGR
Roe, Capt R. G. B. & Mrs I. G. R & IR
Roper, Miss I. M. (1865-1935) IMR
Sandwith, Mrs C. I. (1871-1961) CIS
Sandwith, N. Y. (1901-1965) NYS
Sandwith, Mrs C. I. & N. Y. C & NS
Scase, R. P.
* Scott, Dr L. I. LIS
* Seaward, D. DS
* Shaw, Rev C. E. CES
* Silcocks, Mrs M. A.
* Smith, Dr C. E. D.
* Smith, E. S. ESS
Smith, Dr M. C.
* Spinney, T. G.
Stace, Dr C. A.
* Stafford, Mrs J.
* Stearn, L. F. LFS
Sterne, Mrs G.
Stewart, Mrs O. M. OMS
* Storer, B. BS

LIST OF CONTRIBUTORS

Stribley, Dr D. P.
* Thomas, Miss A.
* Thomas, E.
Thompson, H. S. (1870-1940) HST
Townsend, C. C. CCT
Trapnell, C. G.
* Tulloh, Mrs M. MT
* Twiney, Miss P. PT
* Vince, Miss M. E.
Wade, Dr P. M.
Wallace, T. J. TJW
Ward, H. G. HGW
Watson, Dr W. (1872-1959) WW
Webster, Miss M. McC.
White, Miss A. E. (d. ca. 1945) AEW
White, J. W. (1846-1932) JWW
Whitfield, Miss M.
* Wilkinson, Mrs V. M.
Willis, Prof A. J. AJW
Wilson, Miss K.
* Winchester, Miss A. AW
* Wolseley, Mrs P.

Simplified
Geological Map
of Somerset

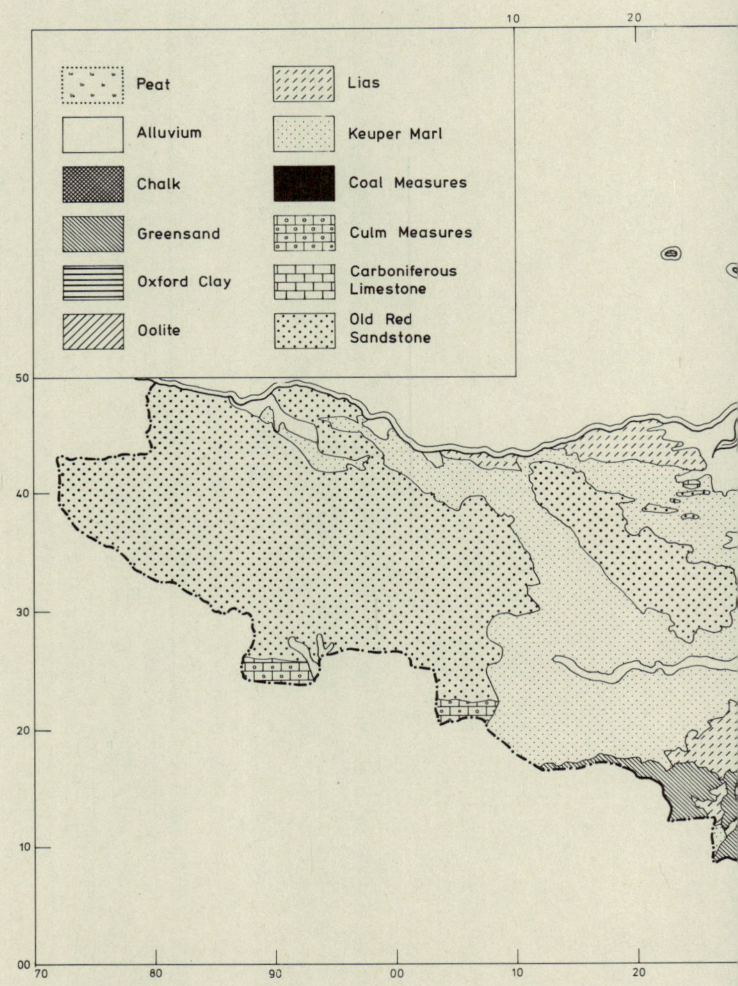

Peat
Alluvium
Chalk
Greensand
Oxford Clay
Oolite
Lias
Keuper Marl
Coal Measures
Culm Measures
Carboniferous Limestone
Old Red Sandstone
10
20
50
40
30
20
10
00
70
80
90
00
10
20

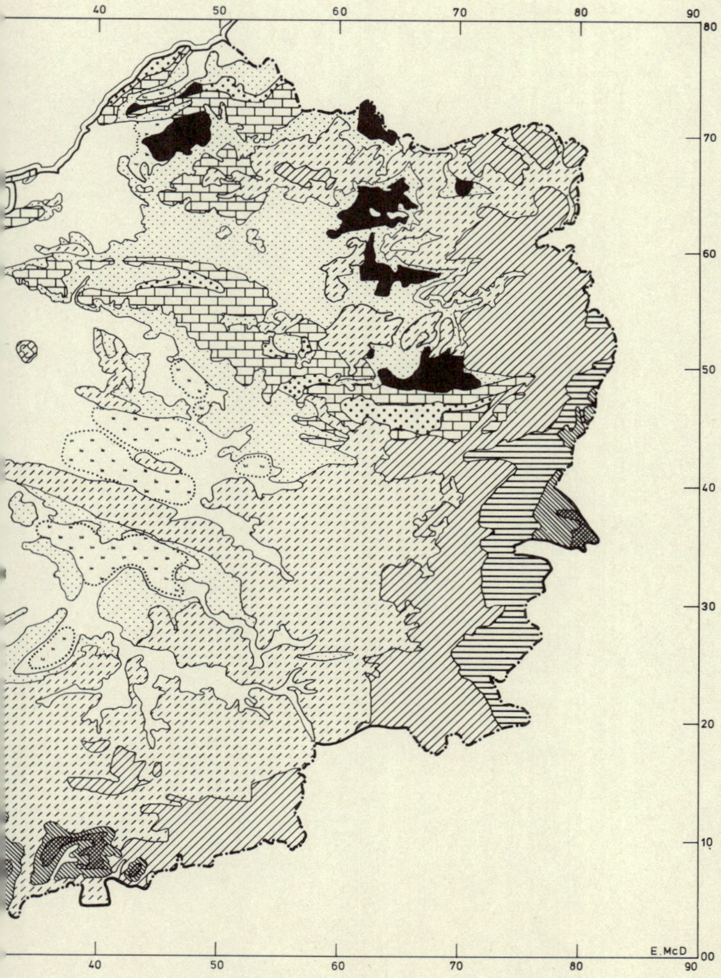
40
50
60
70
80
90
80
70
60
50
40
30
20
10
00
E.McD

Topography of Somerset showing (in red) Vice-County and Botanical district boundaries

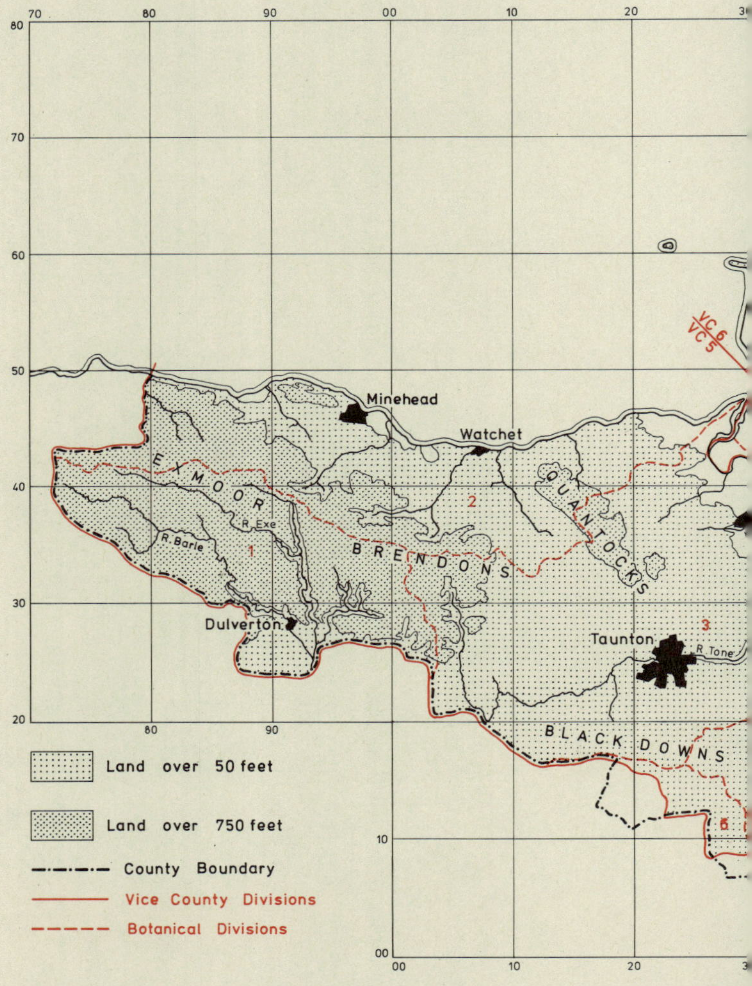

Minehead
Watchet
EXMOOR
R. Exe
R. Barle
BRENDONS
QUANTOCKS
Dulverton
Taunton
R. Tone
BLACK DOWNS
VC 6
VC 5
Land over 50 feet
Land over 750 feet
County Boundary
Vice County Divisions
Botanical Divisions

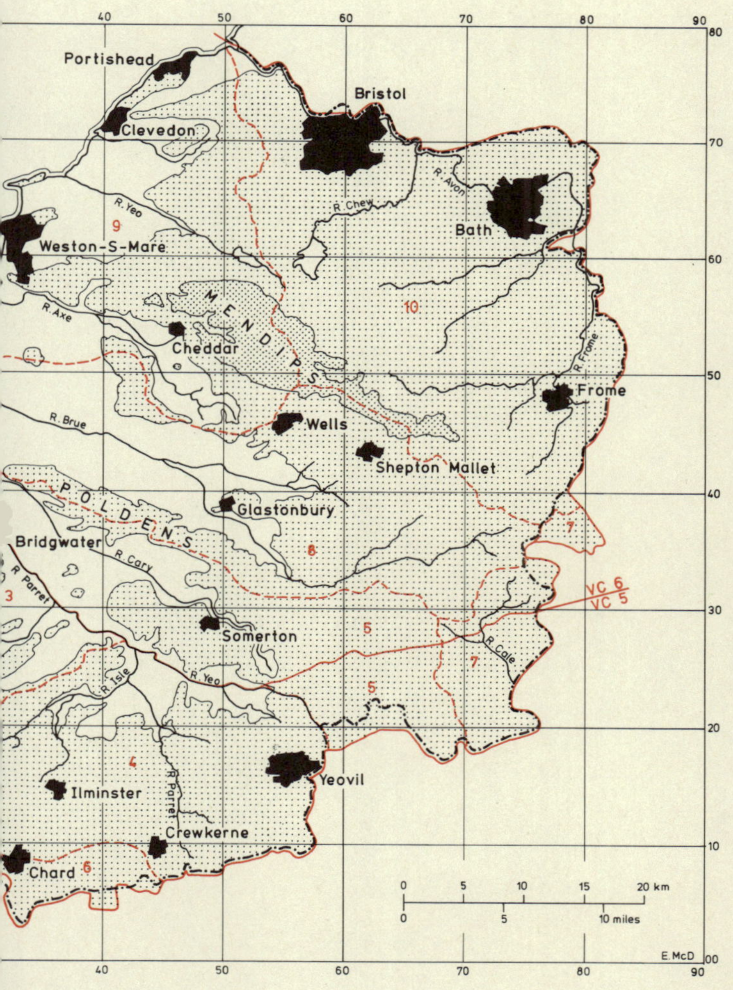

Portishead
Clevedon
Bristol
R. Avon
R. Chew
Bath
R. Yeo
9
Weston-S-Mare
R. Axe
MENDIPS
Cheddar
10
R. Frome
Frome
R. Brue
Wells
Shepton Mallet
POLDENS
Glastonbury
Bridgwater
R. Cary
8
7
R Parret
3
VC 6
VC 5
Somerton
5
R. Cale
7
R. Isle
R. Yeo
5
4
Yeovil
Ilminster
R. Parret
Crewkerne
Chard
6
0 5 10 15 20 km
0 5 10 miles
E. McD
40 50 60 70 80 90
80 70 60 50 40 30 20 10 00

INTRODUCTION

THE HISTORY OF SOMERSET BOTANY

Although William Turner lived at Wells for a time and Somerset was visited by many of the early botanists, none of them left us with a general account of the flora. William Sole (1739-1802), a surgeon and apothecary of Bath, was the first to do this; in 1782 he compiled a Flora in manuscript and although this survived till at least 1856, when it was in the possession of T. B. Flower, it is unfortunately now lost. He later furnished lists of plants with localities for Collinson's *History of Somersetshire* (1791) and Warner's *History of Bath* (1801). Turner and Dillwyn in their *Botanist's Guide* (1805) drew largely on Sole for their Somerset records, sometimes misquoting him so that he has been blamed for errors perpetuated in later works.

C. C. Babington (1808-95) came, with his parents, to live in Bath as a boy of 14 and began to study the local flora. His *Flora Bathoniensis* (1834) and its *Supplement* (1839) give the first complete list we have of the species in a small area of Somerset. Babington quotes records by a number of other botanists, including Dr Davis who supplied the plant list for Warner's *Historical and Descriptive Account of Bath* (1802), as well as his own. Three who contributed notably to the *Supplement* were Dr R. C. Alexander (later Prior) (1809-1902), C. E. Broome (1812-86) and T. B. Flower (1817-99).

H. C. Watson's *New Botanist's Guide* (1835-37) included new Somerset lists supplied by Dr A. Gapper (later Southby) and the Rev J. C. Collins who both lived at that time in the Bridgwater area. Gapper had previously lived at Charlton Adam and a herbarium compiled by him in that area in 1822 was presented to the county Society by the Essex Field Club a hundred years later; the specimens are now some of the earliest Somerset ones at Taunton. The Somersetshire Archaeological and Natural History Society was formed at Taunton in 1849. Among the leading botanists in its early days were the Rev C. P. Parish of Staple Fitzpaine, T. Clark of Bridgwater and Miss Isabella Gifford of Minehead.

This was a time of increasing interest in botany. Two local Floras, E. H. Swete's *Flora Bristoliensis* (1854) and G. St Brody's *Flora of Weston* (1856), add to our knowledge of conditions then. St Brody's work has been criticized as containing many errors but later finds have vindicated many of his records which had been doubted.

In 1855 the Rev L. Jenyns (who later changed his name to Blomefield) founded the Bath Natural History and Antiquarian Field Club with the object of investigating the natural history and antiquities of the neighbourhood. To begin with, its activities were confined to the field, but in 1864 it was expanded and indoor meetings were started. Jenyns was a friend of Darwin and a man of wide interests. His paper "On the Bath Flora", read to the Club in 1866, is valuable in showing the changes which had taken place since Babington's work in 1839. He continued as President of the Club until his death in 1893 and read many papers to them, the last when he was over 90, two years before he died, but none were of local floristic interest and his herbarium does not contain many Somerset specimens.

In 1862 the Bristol Naturalists' Society was founded, a Botanical Section being formed in 1864. The Secretary of the Society and first President of the Botanical Section was Adolph Leipner, who was very keen that the Society should produce a work on the natural history of the Bristol district. When J. W. White came to Bristol in 1874 Leipner persuaded him to undertake the botanical part of this; White's *Flora of the Bristol Coalfield* was published in sections in the Society's *Proceedings* from 1881 to 1886. This book covers quite a large section of North Somerset.

After 1860 little of botanical interest was published in the *Proceedings of the Somerset A. & N. H. S.* until 1875 when Dr H. F. Parsons of Frome presented a paper on the "Flora of the Eastern Borders of Somerset". About 1880 the Rev R. P. Murray of Baltonsborough formed the idea of compiling a *Flora of Somerset* and the resulting work was published in parts in the *Proceedings* of the County Society between 1893 and 1896. Murray drew upon all previously published information, including Watson's *Topographical Botany* (1873-74), and acknowledged the help of several correspondents from other parts of the county, particularly White and his friend, D. Fry, as he apparently did little botanizing in the north himself. This book constitutes the first complete and comprehensive description of the county flora and provides a firm and reliable datum from which to reckon subsequent changes. Murray's work was enthusiastically carried on by the Rev E. S. Marshall, who became Rector of West Monkton, near Taunton, in 1904. He published regular lists of Somerset plant records in the *Journal of Botany* and in 1910 a Botanical Section of the County Society was formed with Marshall as Recorder and W. D. Miller as Secretary. At the same time, although general botanical interest had flagged at Bristol, White and a small circle of friends were energetically working in the north of the county. In 1912 White published his *Flora of Bristol*. Probably the best local Flora ever produced, this is a most entertaining book, full of information which is still valuable to the field botanist 70 years later, though of course there have been a great many changes over the years and some of the localities described have been submerged or transformed by housing, road or other development. The area covered includes a large part of North Somerset, most of Murray's districts 8, 9 and 10. Marshall's *Supplement to the Flora of Somerset,* published by the Somerset A & N. H. S. in 1914, includes a condensed version of the new records in White's book as well as his own and those of other botanists in the western and southern parts of the county. It is sad to relate that while there was all this activity at Taunton and Bristol, the Bath Field Club had declined and did not long survive the death of its founder in 1893. In 1908 the drastic step of admitting ladies was agreed in the hope of reviving interest but by 1912 the Club had ceased to exist.

From this time onwards botanical records have been continuously maintained by both the main societies operating in Somerset. At Taunton Marshall continued to publish his annual lists of Somerset records in the *Journal of Botany* and a selection of the more important ones in the *Proceedings of the Somerset A. & N. H. S.* until his untimely death in 1919. W. D. Miller then took over the recording, compiling a card index, which he added to over the next 14 years. On his death Dr W. Watson continued the work until 1952 when he handed over to A. D. Hallam. During these years a great deal of information was collected; besides the Recorders named, leading contributors were Miss I. M. Roper, Dr. H. Downes, H. S. Thompson, N. G. Hadden, F. K. Makins and E. J. Hamlin.

In 1918 White published "Notes Supplemental to the Flora of Bristol" in the *Journal of Botany* and from 1917 to 1930 he contributed notes and lists of records to the *Proceedings of the Bristol Naturalists' Society* under the title "Bristol Botany in 19——". In 1909 Mrs C. I. Sandwith came to Bristol and she and her son, N. Y. Sandwith, were soon contributing many records to "Bristol Botany". In 1933 Mrs Sandwith's paper "The Adventive Flora of the Port of Bristol" was published in the *Report of the Botanical Society and Exchange Club of the British Isles for 1932.* White died in 1932 and in 1935 Mrs Sandwith revived the "Bristol Botany" notes and continued to write them (from 1947 jointly with her son) for the next 26 years. N. Y. Sandwith carried them on until his sudden death in 1965. "Bristol Botany" has been continued since by A. J. Willis, to whom was left the card index meticulously kept by Sandwith for many years. Among the botanists who also worked in the Bristol area during the 50 years after the publication of the *Flora of Bristol* may be mentioned C. Bucknall, Mrs E. S. Gregory, Miss M. A. G. Livett, Miss I. M. Roper, H. S. Thompson, the Rev E. Ellman, H. J. Gibbons, T. H. Green and I. W. Evans.

INTRODUCTION

The outbreak of war in 1939 and the move of a large part of the Admiralty to Bath was the occasion for the revival of organized botanical activity there with the founding of the Bath Natural History Society. Miss E. H. Stevenson, R. P. Scase and D. E. Coombe were the leading botanists in its early days.

Field botany in Somerset was stimulated by the Distribution Maps Scheme started by the Botanical Society of the British Isles in 1954. This was supported by all the Natural History societies in the county and many members took part. The scheme suggested the idea of similar surveys on a more detailed scale. P. F. Hunt, who then lived at Frome, became interested in the idea of a County Survey but on moving to Middlesex felt unable to undertake the organization himself. In 1965 he persuaded the present author to do so and the following year the Survey was formally launched at a B. S. B. I. meeting at Crewkerne. The method chosen was based on the B. S. B. I. Maps Scheme but using a 2 kilometre square (tetrad) as the recording unit. Record cards were printed with the abbreviated names of the commoner species on one side and the other blank for recording the other species found, with grid references, habitat and other supporting details. The records received were plotted on map cards on a scale of 20 km to 1 in. All records from 1960 onwards were admitted to the Survey; the date was chosen so that information collected by R. M. Burton in a detailed survey based on Weston-super-Mare and from another by my wife and myself in the Bath area could be included. A card index of all published and other former records which could be traced was made, and this information also was plotted on the species map cards using special symbols so that changes in the distribution patterns were shown up. The species maps and the tetrad record cards together constitute a complete record of the Survey over the period 1960-80.

Somerset was rather slow in entering the nature conservation field and it was not until 1964 that the Somerset Trust for Nature Conservation was formed. Since then its activities have expanded year by year, Nature Reserves have been established and the Trust has co-operated with the Natural History societies to oppose threats to the County's wild life and its habitats. Unfortunately this action was too late to prevent the devastation of the Somerset peat moors by large-scale peat extraction, planning permission for which was given in 1949-50. Even the National Nature Reserve of Shapwick Heath (established in 1961) is largely scheduled to be excavated and it appears that only a few corners of the once extensive peat moors can be salvaged. Other parts of the Somerset Wetlands are under threat by drainage schemes and the Carboniferous Limestone of Mendip is much in demand for road-making but it is hoped that public awareness was awakened in time to prevent too drastic an extension of quarrying.

TOPOGRAPHY

The ancient county of Somerset is some seventy miles from west to east and fifty from north to south, and has within its limits a very wide and varied range of scenery and habitats. In the west, Exmoor and the other uplands are a part of moorland Britain and contrast sharply with the limestone downs of Mendip and the hills round Bath and along the eastern border. In the centre of the county are the extensive Somerset Levels, once flooded by the sea but now drained marshland mainly below the 25-foot contour and containing the remnants of the "peat moors", a raised bog area which is rapidly being exploited for peat. Along the north-west the Bristol Channel forms the boundary and provides a variety of maritime habitats, cliffs, shingle, sand dunes and salt marshes. The highest point is Dunkery Beacon, at 1707 feet, on Exmoor. Much of the western moorland is over 1000 feet in altitude. Elsewhere the highest part of Mendip is just over 1000 feet but other hills do not reach this height though the contrast with the Levels makes even quite low hills look higher than they are.

The central lowlands are drained by the Rivers Parrett, Cary, Brue, Axe and Yeo. In the west the Rivers Exe and Barle flow south from Exmoor. In the south another Axe forms part of the Dorset border and the Cale, a tributary of the Dorset Stour, drains the south-eastern corner of the county. In the north the lower course of the Avon divides the county from Gloucestershire.

GEOLOGY AND SOILS

With some minor exceptions the oldest rocks in Somerset are the Devonian sandstones, slates and shales which make up Exmoor and the Brendon and Quantock Hills. These give rise to poor acid soils on which the natural vegetation is moorland. In a few places there are small areas of limestone, usually marked by quarries, where patches of alkaline soil may be found. Completely contrasting both in scenery and flora is the Carboniferous Limestone which forms the Mendip Hills and the outcrops further north of Worle Hill, Sand Point, Broadfield Down and the hills between Clevedon and Bristol which are cut through by the Avon Gorge. In this area alkaline soils predominate but the underlying Old Red Sandstone is exposed at the top of the Mendip anticline, in the Avon Gorge and in the Clevedon to Portishead ridge, and acid soils occur in these places. Locally acid conditions also develop on the top of limestone hills (Limestone Heath) due to the build up of humus from dead vegetation and leaching of the soluble limestone by rain.

In the basin north of Mendip lie the Coal Measures which form the Somerset part of the Bristol Coal Field. Most of this formation is buried beneath the succeeding Trias and Lias but there are some outcrops of Pennant Sandstone and the Coal reaches the surface in places. The coal seams are thin and much contorted and faulted and the coal mines are no longer worked. The Coal Measures provide small areas of acid soils in a region of predominantly alkaline ones.

The Armorican earth movements which caused the folding of the Devonian and Carboniferous strata and the formation of Exmoor and the Mendip and other hills referred to, were succeeded by the Triassic period. New Red sandstones and mudstones were laid down in a sea which filled the valleys. The soils derived from this formation are fertile marls and clays. In the west they line the Vales of Porlock and Taunton Deane and provide a contrast to the poor acid soils of the adjoining moorland. They occupy a large part of the basin north of Mendip and here the soil is more calcareous because of the alkaline drainage water from the surrounding limestone hills.

The oldest of the Jurassic strata is the Lias, and rocks of this age occupy a large area of central Somerset. Largely clays and shales, there are also beds of limestone which crop out as low hills and which have been much quarried for building stone. There is a narrow band of Lias north of Quantock as far west as Blue Anchor but the main formation lies south and east of Taunton. It is exposed on the north face of the Blackdown Hills, forms the ridge between West Hatch and Langport and the higher ground eastwards as far as Castle Cary, and the Polden Hills. The Oolites stretch from the southern end of the Cotswolds north of Bath down the eastern side of the county to the Dorset border and south of Yeovil to Crewkerne; they are limestones and clays and form calcareous soils. Between the Oolite and the Cretaceous strata east of Frome and Bruton there is a belt of Oxford Clay which makes a neutral, heavy soil.

Cretaceous rocks form the Blackdown Hills on the southern border of the county and the high ground on the eastern border between Penselwood and Gare Hill. In Somerset they consist mainly of Gault Clay and Greensand. The Greensand is acid and has a very distinctive flora, particularly round the flushes at the junction with the underlying Jurassic clays. Sharply contrasting is the Chalk itself, of which there are small outcrops near Chard, and at Whitesheet Hill and Long Knoll which are now in the administrative county of Wiltshire, although just within the eastern boundary of old Somerset.

In geologically recent times the central lowlands were extensive marshes in which beds of peat built up. Later these were largely covered by alluvium brought down by the rivers and an incursion by the sea washed away the western part of the peat and laid down a broad belt of estuarine mud. The peat is now only exposed in the remains of the raised bog forming the "peat moors". In other parts of the Levels the thickness of the alluvium varies greatly; in some places the peat is near the surface and is exposed in the drainage ditches. The characteristics of the soil depend on the parent material of the surface deposits as well as on their thickness. The high water table and mainly alkaline drainage water make most of the Levels into the second largest fenland in England. In this area the field boundaries are defined by ditches and the main drainage channels are known as "rhines" or "reenes". Compared with the main inland parts of the Levels the belt of estuarine alluvium, where there is no underlying peat, is floristically dull. The term "Somerset Levels" is usually confined to the area south of the Mendips but the basin to the north is just the same. The Gordano Valley between Portishead and Bristol is a miniature version of the Levels.

THE BOTANICAL DISTRICTS

Murray in his *Flora of Somerset* divided the county into ten districts founded on the river basins. Although this system of sub-division is not ideal botanically, and the district boundaries are not marked on Ordnance Survey maps, the watersheds can fairly readily be made out. As the county records have been continued using Murray's districts ever since his day, the same system is used in this work. As can be seen from the map, districts 1, 2, 3, 4 and 6 are in V.C.5, South Somerset, and 8, 9 and 10 in V.C.6, North Somerset. Districts 5 and 7 are divided by the vice-county boundary, most of 5 and part of 7 lying to the north and the rest to the south.

District 1 — (DULVERTON) — The basin of the Exe and its tributaries the Barle, Quarme, Haddeo and Batherm. This district covers the southern part of Exmoor and the Brendon Hills and lies almost wholly on Devonian strata, so the soils are poor and acid. There are a few small patches of limestone which have been quarried at Newland near Exford and west of Wheddon Cross and some even smaller ones near Dulverton, but these are not large enough to have much influence on the flora and most calcicolous species are absent from the district. Its main botanical interest is on the high moors round Simonsbath, Exford, Withypool, Winsford and north of Dulverton. There is a different flora in the sheltered and partly wooded lower river valleys near Dulverton.

District 2 — (MINEHEAD) — The whole of Somerset west of the Parrett which is drained northwards into the Bristol Channel. There are no large rivers in this district but a number of small streams draining the northern slopes of Exmoor and the Brendon and Quantock Hills. As in district 1 the main physical features are the Devonian uplands with a similar flora, but the Vale of Porlock which separates North Hill, Minehead, from the main block of Exmoor, and the vale between the Brendons and Quantocks are floored with Triassic marls giving deeper and richer soils. A narrow belt of Lias stretching from Blue Anchor round the north end of Quantock provides a home for many calcicolous species missing from district 1. The coast from the Devon border to Porlock and again from Bossington to Minehead is of steep cliffs with rather inaccessible beaches. The western cliff tops are heavily wooded and though there are vestiges of the indigenous flora, including the very rare *Sorbus vexans* and *S. subcuneata,* most of the trees were introduced in private plantations last century. The shingle and marshland between Porlock Weir and Hurlstone Point provide a habitat for many coastal and maritime species. Further east the sand dunes of Minehead Warren and Dunster Beach are of great botanical interest. From Blue Anchor to Hinkley Point there are low cliffs and muddy, shingly beaches. The shore of Bridgwater Bay between Stolford and Stert Point has extensive mudflats with shingle higher up, backed by a narrow belt of salt marsh and sandy commons. The district has a range of habitats as varied as any in the county and consequently a very rich flora.

District 3 — (TAUNTON) — The southern half of the basin of the Parrett from Langport to the sea. The southern slope of the Brendons and the greater part of the Quantocks occupy the north-western part of the district. The Devonian sandstone carries a moorland flora similar to that in districts 1 and 2. There are no blanket bogs as on Exmoor but the combes and valleys, particularly on the eastern slope of Quantock, are interesting in spite of the extensive conifer plantations. The Vale of Taunton Deane and the lower ground between the Quantocks and the Parrett are Triassic marls which provide good farmland and orchards. Some interesting small fens have developed in marshy moorland between Wiveliscombe and Milverton and the more acid common of Langford Heathfield is unique, with some species not found elsewhere in the county. The Tone valley and the remains of the Grand Western Canal west of Taunton have a good flora. South of Wellington and Taunton the district is bounded by the Blackdown Hills and the Lias ridge between Pitminster and Langport with outliers near Stoke St Mary and the low hills from Thornfalcon to Stoke St Gregory. The Lias also crops out on the lower slopes of the Blackdowns. It has bands of limestone and a typical calcareous flora. There is a small outlier of Carboniferous Limestone at Cannington in the north of the district. The Greensand and Gault clays overlie the Lias in the Blackdown Hills and there are many marshy flushes along the northern slope at the junction of the Greensand and the Lias clays. The Bridgwater and Taunton Canal, now disused except by fishermen, has a good aquatic flora. The Parrett and the lower reaches of the Tone flow through alluvium forming the western end of the Somerset Levels. The Moors, as they are called, are liable to flooding and are intersected by drainage ditches and rhines with many aquatic and marsh species. The most important part of this fenland is West Sedge Moor, where peat occupies the centre of the moor with only a very thin covering of alluvium. The moor is as yet undrained and much of the ground is too soft to support heavy agricultural machinery. Proposals to lower the water table by pumping, as has been done in other parts of the Levels, have been strongly opposed by nature conservation interests as they would have a devastating effect on the plant and bird life of the moor. The lower part of the Parrett estuary below Bridgwater is lined by mudflats and salt marshes.

District 4 — (ILMINSTER and YEOVIL) — The basin of the Parrett above Langport and the southern half of that of its tributary the Yeo. The greater part of this district lies on Liassic strata, limestone bands and calcareous clay in the north-western ridge separating it from West Sedge Moor, giving way to lighter and more sandy soil in the southern part where there are extensive tracts of the Yeovil Sands. Harder sandstone crops out at Ham Hill where it has been much quarried for building. From Crewkerne eastwards Oolite and Fuller's Earth clay overlie the Lias. In the south-west the district is bounded by the northern slope of the Blackdown Hills, where Greensand and Gault clay carry a similar flora to that of the same formations in district 3 further west. Alluvial Levels come inland as far as Kingsbury Episcopi and Ilchester. The moors round Muchelney are particularly subject to flooding because the narrow gap in the Lias hills at Langport restricts the flow away of flood water.

District 5 — (SOMERTON) — The northern half of the basin of the Parrett and Yeo. The northern boundary is formed by the Polden Hills, a low Lias ridge running from Puriton to Keinton Mandeville and on to Castle Cary. A larger mass of Lias stretching east from Langport and around Somerton occupies much of the district and has been much used for building in this area. On the eastern border near South Cadbury the hills are of Oolite. An important tributary of the Parrett is the Cary. The lower course of this river has been artificially straightened, widened and deepened and is now called the King's Sedgemoor Drain. Peat underlies the alluvium of Somerton Moor and King's Sedge Moor and adds interest to the flora. The Levels are bordered by Triassic clays on the lower slopes of the surrounding hills. From Othery to Weston Zoyland recent deposits of sand and gravel make a low "island" of slightly higher ground. There are practically no acid soils in this district and many calcifuge species are absent.

District 6 — (AXE) — The southern flank of the Blackdown Hills and the ridge east of Chard drained southwards by the Culm, Otter and Axe into Devonshire and Dorset. This is a small narrow district but the Lias, Greensand, Gault clay and small exposures of Chalk near Chard give a good range of soils from acid to alkaline and quite a varied flora, but the total number of species present is smaller than in most other districts. Chard Common was once a particularly interesting marshy area but although a few good patches remain, most of it has been drained and cultivated or built over.

District 7 — (WINCANTON) — The south-eastern corner of the county draining southwards into Dorset and Wiltshire. Another small district the larger part of which south of Wincanton is drained by the Cale, a tributary of the Dorset Stour. It is mainly Oolite and other middle and upper Jurassic strata but round Penselwood they are overlain by Greensand. Further north round Kilmington is an area of Chalk, part of the Wiltshire Wylye basin. As in district 6 the number of species present is limited by the small area.

District 8 — (GLASTONBURY) — The basin of the Brue extending from the Bristol Channel at Burnham to the Wiltshire border east of Bruton. The northern boundary is the low Lias ridge from Brent Knoll and Wedmore to Wells and then runs along the south flank of Mendip. The district contains the full range of geological strata from the Old Red Sandstone, exposed on the Mendip crest at Pen Hill and Maesbury, to the Greensand east of Bruton, the alluvium of the Levels and the famous "peat moors". The flora is rich and varied. Very little of the raised bog remains on the peat moors and species formerly frequent there have disappeared, but an interesting peaty fenland remains. The peat was formerly cut by hand but is now being rapidly dug out by machinery. Proposals have been made to flood the diggings when extraction has ceased and make a modern version of the Norfolk Broads; whatever happens the character of the area and consequently its flora is bound to be greatly changed.

District 9 — (MENDIP) — The whole of the area draining into the Bristol Channel by streams between the mouths of the Brue and the Avon. The largest rivers are the Axe, Yeo and Kenn. Carboniferous Limestone is the main feature of the district. It forms the main ridges of Mendip from Wells to the coast at Brean Down and the island of Steep Holm, with outlying outcrops at Worle Hill and Sand Point, and the upland of Broadfield Down. It also forms the ridges from Clevedon to Portishead and to Failand. The underlying Old Red Sandstone is exposed in the anticline at Blackdown on Mendip, in the cliffs between Clevedon and Portishead and above Clapton-in-Gordano. Triassic clays cover much of the inland basins between the limestone ridges, but the Coal Measures surface near Nailsea. The Axe valley and the lowlands nearer the sea are fenny Levels similar to those in other districts. Peaty areas occur below Cheddar and Axbridge and on Kenn, Nailsea and Tickenham Moors and in the Gordano Valley. Many plant species are confined to the Carboniferous Limestone, not only in the famed Cheddar Gorge and other combes but also on the rocky south-facing downland of Brean Down, Uphill, Bleadon and Crook Peak. The other special feature of the district is the sand dune system at Berrow and Brean; there are smaller dune areas at Uphill and Kewstoke. Areas of salt marsh lie behind the muddy parts of the coast and line the river estuaries. The Blagdon Lake, a reservoir formed by a dam on the Yeo, has a good aquatic and marsh flora. The wholly artificial Cheddar Reservoir is floristically dull in comparison.

District 10 — (BRISTOL and BATH) — The basin of the Avon with its tributaries the Frome, Midford and By brooks and the Chew. This is the largest of the districts and stretches from the Carboniferous Limestone of Mendip in the south to the Gloucestershire border which from the sea to above Keynsham is the course of the Avon itself. A small area round Bath on the right bank of the river is included in Somerset. The eastern boundary is the Wiltshire border.

The hills round Bath and south to Frome are of Oolite and have a rich calcareous flora. South of Frome there is a belt of Oxford Clay, Gault and Greensand between the Oolite and the border. The lower ground west of Bath is a basin formed by a syncline of the Carboniferous Limestone bordered by Lias and filled by Triassic Clay. The Coal Measures crop out at Holcombe, Temple Cloud, Clutton and Keynsham and their sandstones and shales give rise to the main areas of acid soil in the district. The Coal has been worked at many places but the last pit near Radstock closed recently. The many spoil heaps, except a few of the most recent ones, are well covered by vegetation which gives the area a very different appearance from most other coalfields. In the west of the district the Chew Valley Lake is a fairly recently made reservoir with a good aquatic flora. Lower down near Publow and Compton Dando the Chew runs through Coal Measures sandstone with surrounding Lias hills, and the variety of soils makes it a very interesting area. In the lowest part of its course the Avon runs through the Avon Gorge, the Somerset side of which, backed by Leigh Woods, is every bit as interesting as the better-known Gloucestershire side. Below the Carboniferous Limestone the Old Red Sandstone is also exposed here. Below the Gorge mudflats and salt marshes border the estuary and this, with a narrow belt by the river higher upstream, provides the only habitat for maritime plants in the district.

THE FLORA

Area surveyed

For biological recording purposes Somerset is divided into two "vice-counties", V.C.5 (South Somerset) and V.C.6 (North Somerset). Their boundaries were established in 1852 by H. C. Watson. Since then there have been a number of administrative changes by which whole parishes as well as smaller areas have been transferred from and to adjacent counties. Parishes lost are Seaborough, Trent, Sandford Orcas and Poyntington to Dorset and Gasper, Kilmington and Yarnfield to Wilts. Parishes added are Churchstanton from Devon (V.C.3) and Wambrook from Dorset (V.C.9). The survey included these added parishes as well as the whole of V.C.s 5 and 6. In 1974 a large part of North Somerset was taken to form part of the new county of Avon. The majority of the inhabitants of this area as well as of the ancient County Boroughs of Bristol and Bath were against this change and whether the new Local Government boundaries prove permanent remains to be seen, but no notice of this administrative upheaval has been taken in the survey.

Sequence and Nomenclature

For the **Pteridophyta** nomenclature and order of the genera follow that of the *Atlas of Ferns of the British Isles* (ed. A. C. Jermy et al., 1978). For the **Spermatophyta** the sequence of families, genera and species is that of the *List of British Vascular Plants* (ed. J. E. Dandy, 1958) and the nomenclature is that of *Flora Europaea* (ed. T. G. Tutin et al., 1964-80) with a few amendments. Synonyms given are those used by Murray (1896), White (1912), Clapham, Tutin & Warburg (1952) or Dandy (1958). English names, with a few exceptions, are taken from Dony, Perring & Rob's *English Names of Wild Flowers* (1974).

Habitat, Frequency and Distribution

Species believed to be extinct in the county are marked with a dagger (†). Where plants are not believed to be native in Somerset the status, and whether naturalized or casual, is indicated. All species recorded during the survey are listed with the exception of a few garden escapes and planted trees and shrubs which have only been recorded once or twice. Casuals which have been recorded in the county at some time before 1960 but not since are listed in the Appendix with the dates of their last records. Habitats are those in which the plant has been found in the county in order of frequency. Frequency is described in the usual terms (common, rare, etc.,) and has been assessed from the species maps prepared during the survey. Distribution has also been taken from the maps and refers to the situation in the period 1960-80. For the more frequent species distribution is given in general terms using the main topographical features and the botanical districts. Where a species has not been recorded in all districts this is indicated. The terms "recent" and "formerly" in this context mean since or before 1960 respectively.

Records

Where a plant is rare in the county or a district, localized records are given but not in great detail for obvious reasons. Place names are followed by the number of the botanical district in which they are situated. Unless of special interest the name of the original finder is not stated where records are of long standing but where new records have been made during the survey the date and finder's name or initials are given. Where reference is made to herbarium specimens the following abbreviations are used for the herbaria of the Institutions indicated:-

BM	British Museum (Natural History), Dept. of Botany.
K	Royal Botanic Gardens, Kew.
LDS	Leeds University, Dept. of Botany.
LIV	Merseyside County Museums, Liverpool.
LTR	Leicester University, Dept. of Botany.
MANCH	Manchester University, Dept. of Botany.
OXF	Oxford University, Dept. of Botany.
TTN	Somerset Archaeological & Natural History Society, Taunton.

PTERIDOPHYTA

LYCOPODIACEAE

LYCOPODIUM L.

Lycopodium clavatum L. Stag's-horn Clubmoss

A rare and decreasing plant of acid moorland now only to be found in a few spots on Exmoor. It was formerly recorded from the Brendon, Quantock and Blackdown Hills, Blackdown on Mendip and from near Clevedon (9), and it may still survive in some of these areas. A few young plants were found in 1970 near Yeovil (4) (CJC) where they had apparently been introduced with conifers in cleared and replanted woodland.

LYCOPODIELLA Holub

†Lycopodiella inundata (L.) Holub Marsh Clubmoss

Lycopodium inundatum L.; *Lepidotis inundata* (L.) C. Börner

Recorded from Exmoor and the Blackdown Hills about 1850, but not seen since.

HUPERZIA Bernh.

Huperzia selago (L.) Bernh. ex Schrank & Mart. Fir Clubmoss

Lycopodium selago L.

Only recent record is near Pinkworthy Pond (1), 1970 (JGK). Formerly recorded from acid moorland in other parts of Exmoor, the Brendon and Blackdown Hills, Blackdown on Mendip and from near Clevedon (9). Not yet found on Quantock; the plant from Will's Neck referred to in Marshall (1914) was later identified as *Lycopodium clavatum.*

DIPHASIASTRUM Holub

†Diphasiastrum alpinum (L.) Holub Alpine Clubmoss

Lycopodium alpinum L.; *Diphasium alpinum* (L.) Rothm.

Formerly recorded from a few localities on Exmoor but not seen in the county since 1927 (NGH).

EQUISETACEAE

EQUISETUM L.

†Equisetum hyemale L. Rough Horsetail

Formerly on some damp sandy ground near the railway station at Weston-super-Mare (9), where it was last recorded in 1933 (WDM). The exact date of its disappearance is not known but the site was then under serious threat by development. An 1802 record for the canal bank at Bath was never subsequently confirmed.

Equisetum variegatum Schleich. ex Weber & Mohr Variegated Horsetail

Now known only from a dune slack at Berrow (9), but it also grew at Weston-super-Mare in the same place as *E. hyemale*. It was last recorded there in 1951 (EJH), by which time the other species had already gone.

Equisetum fluviatile L. Water Horsetail

E. limosum L.

Rather common in peaty ditches of the lowlands from the Gordano valley to the moors south of Langport, but less so in the alluvial clay area near the coast. Elsewhere it is not uncommon in shallow water and marshy places by ponds, streams and rivers.

Equisetum arvense L. Field Horsetail

Very common almost everywhere, but less so on the higher ground of Exmoor, the Quantocks and Mendips.

× **fluviatile** (*E.* × *litorale* Kühlew. ex Rupr.). Known only from the boggy valley below Shipham (9) and from Chard Common (6). Not seen recently in Marshall's 1905 locality near Dulverton (1).

Equisetum sylvaticum L. Wood Horsetail

Wet wooded slopes of the Blackdown Hills in districts 3, 4 and 6, and again on the eastern border of the county in districts 7, 8 and 10, from Gasper to Witham and Selwood east of Frome, where springs mark the junction of the Greensand with underlying clays. A small patch near Cranmore (10) where the Forest Marble overlies Fuller's Earth Clay, was first reported in 1955 (NYS). On Exmoor it is a rare plant of the upper valleys of the Exe and Barle near Simonsbath (1). Only once recorded (1921, NGH) from district 2 on Exmoor. Unknown elsewhere. A very old record from near Bath (10) has never been confirmed.

EQUISETACEAE

Equisetum palustre L. Marsh Horsetail

Rather too common in damp meadows and marshes. A nuisance to dairy farmers as it causes scouring and low milk yields when eaten by cows.

Equisetum telmateia Ehrh. Great Horsetail

E. maximum auct.

Rather common in wet woodlands and shady places over much of the county but very rare west of Wiveliscombe (3) and Blue Anchor (2). The only recent Exmoor record is from the Exe valley above Exford (1), 1973 (JGK), but it was seen by Marshall in 1918 near Simonsbath (1). Further north the only record is from Selworthy (2), 1975 (CJG).

OPHIOGLOSSACEAE

BOTRYCHIUM SW.

Botrychium lunaria (L.) Sw. Moonwort

On heaths and rough hill pastures. Now very rare and only seen recently near Exford (2), 1980 (Miss M. Davidson), at two spots on Mendip and on Brean Down (9). Recorded previously from all districts except 5 and 7 and always rare, it has probably been much affected by ploughing of marginal land for re-seeding and afforestation.

OPHIOGLOSSUM L.

Ophioglossum vulgatum L. Adder's Tongue

Rather rare in damp pastures and woodland rides. More frequent on the Carboniferous and Jurassic limestones than elsewhere. Decreasing on lower ground and not seen for many years in districts 1 or 2.

OSMUNDACEAE

OSMUNDA L.

Osmunda regalis L. Royal Fern

Still fairly plentiful on the peat moors between Shapwick and Glastonbury (8), but it has gone from old localities on the peat of the Gordano valley and Kenn Moor (9). In one or two places on the Blackdown Hills (6) and one near Chard (4) where it was formerly more frequent. This plant has suffered from fern collectors, and survives as an introduction in the grounds of several large houses in the county.

ADIANTACEAE

Cryptogramma R. Br.

†**Cryptogramma crispa** (L.) R. Br. ex Hook. Parsley Fern

First recorded by N. Ward in 1840 as growing very sparingly on a stone wall near Simonsbath (1), it was lost for many years until refound in 1956 by Mrs C. M. A. Cadell on scree in the same area. It was still there early in 1976 but appears to have succumbed to the severe drought of that summer as it has not been seen since despite careful search by several botanists.

Adiantum L.

Adiantum capillus-veneris L. Maidenhair Fern

Introduced in Somerset but long established on old walls at Wrington (9) and in the Bath area (10). Also very fine on rocks in a disused railway cutting near Wells (9), 1979 (RMP).

HYMENOPHYLLACEAE

Hymenophyllum Sm.

Hymenophyllum tunbrigense (L.) Sm. Tunbridge Filmy Fern

Only on damp shady rocks near Holford and Porlock (2). The record from Shepton Mallet by Sole quoted in the *Botanist's Guide* (1805) has never since been confirmed.

POLYPODIACEAE

Polypodium L.

Polypodium vulgare L. sens. lat. Polypody

Very common on rocks, walls and old trees. The aggregate is now usually divided into three species, **P. vulgare** sens. strict., **P. interjectum** Shivas and **P. australe** Fée. The distribution of these in Somerset is imperfectly known because of the difficulty of accurate identification in the field. *P. vulgare* and *P. interjectum* seem to be widely distributed in the county. *P. australe* is confined to the limestone areas. Hybrids between any two of these species are known and may occur occasionally.

DENNSTAEDTIACEAE

PTERIDIUM Scop.

Pteridium aquilinum (L.) Kuhn — Bracken

Pteris aquilina L.

Very common on moors, rough grassland and in woods over most of the county. Not common in the lowlands except on the peat moors.

THELYPTERIDACEAE

THELYPTERIS Schmidel

Thelypteris thelypteroides Michx ssp. **glabra** Holub — Marsh Fern

T. palustris Schott; *Lastraea thelypteris* (L.) Bory

Only on the peat moor (8) and there plentiful in marshy places. Formerly in two localities near Clevedon (9) but not recorded there since 1915.

PHEGOPTERIS (C. Presl) Fée

Phegopteris connectilis (Michx) Watt — Beech Fern

P. polypodiodes Fée; *Thelypteris phegopteris* (L.) Slosson; *Polypodium phegopteris* L.

Only in a few places near Simonsbath (1), on the Devon border near Wambrook (6), 1975 (T. J. Wallace), and in Leigh Woods (10) where it was first found in 1957 (G. W. Garlick) but seems to have disappeared by 1977 (PJMN). Formerly near Wells (8) but not reported there since 1884. It is said to have been planted near Alfred's Tower (8) about 1830 but has long since gone.

OREOPTERIS Holub

Oreopteris limbosperma (All.) Holub — Lemon-scented Fern

Thelypteris limbosperma (All.) H. P. Fuchs; *T. oreopteris* (Ehrh.) Slosson; *Lastraea oreopteris* (Ehrh.) Bory

Rather common by streams and in marshy woodland on Exmoor, and on the Quantock and Blackdown Hills. Outside this area it occurs now only at a few spots on Mendip and on the Wiltshire border between Kingsettle Hill (8) and Witham (10), though formerly it was more widespread. Not seen recently in district 7 and never yet in 5.

ASPLENIACEAE

Asplenium L.

Asplenium scolopendrium L. Hart's-tongue

Phyllitis scolopendrium (L.) Newm.; *Scolopendrium vulgare* Sm.

Very common in shady woods and lanes almost everywhere in the county.

Asplenium adiantum-nigrum L. Black Spleenwort

Common on hedgebanks, rocks and walls in the west and south but less so in the east. Rather scarce in the lowlands and in the north, where it is mainly found on the stonework of bridges and walls, particularly by railways.

†**Asplenium billotii** F. W. Schultz Lanceolate Spleenwort

A. obovatum auct. angl. non Viv.; *A. lanceolatum* Huds. pro parte.

A specimen from Selworthy (2), ca. 1850 (Miss I. Gifford), is in the Herbarium at Taunton and unconfirmed reports from North Somerset date from about the same time.

Asplenium marinum L. Sea Spleenwort

In rock crevices and caves by the sea, usually within reach of the waves. Rare and only in a few spots on the coast between the county border and Minehead (2), at Weston-super-Mare and Sand Point and between Clevedon and Portishead (9). Also on Steep Holm. Not seen recently at Brean Down (9) but it may well survive there.

Asplenium trichomanes L. Maidenhair Spleenwort

Very common on rocks and walls. Practically all our records belong to ssp. **quadrivalens** D. E. Meyer emend. Lovis, but the calcifuge ssp. **trichomanes** may occur near the border in West Somerset (2).

Asplenium ruta-muraria L. Wall-rue

Very common on old walls and other stonework, but less so on rocks.

†**Asplenium septentrionale** (L.) Hoffm. Forked Spleenwort

Discovered about 1840 on loose stonework near Porlock and Oare (2). It was then plentiful but in *British Ferns* (1854) Newman relates that when these localities became known a fern collector named Potter brought away "hundreds, or perhaps thousands, of roots for sale". It is not surprising that by 1893 it had become very rare and the exact place in which it still survived was kept secret. In 1939 only one plant was seen (NGH) since when there have been no definite reports. A 1726 record by Dillenius for Cheddar (9) was probably an error as the plant is a calcifuge.

× **trichomanes** sens strict. (*A × alternifolium* Wulfen; *A. × breynii* auct.; *A. germanicum* auct.). Found with *A. septentrionale* and specimens from both original localities are in **K,** but there are no later reports of its having been seen.

Asplenium ceterach L. Rusty-back

Ceterach officinarum DC.

Very common on rocks and walls, especially in the limestone districts of the north and east, less so in the south and becoming scarce in the extreme west.

ATHYRIACEAE

ATHYRIUM Roth

Athyrium filix-femina (L.) Roth Lady Fern

Common in damp woods and marshes by streams over most of the county. Scarce in the lowlands except on the peat moors.

GYMNOCARPIUM Newm.

Gymnocarpium dryopteris (L.) Newm. Oak Fern

Thelypteris dryopteris (L.) Slosson; *Polypodium dryopteris* L.; *Phegopteris dryopteris* (L.) Fée

Very rare and now only known in Murray's locality near Landacre Bridge (1). It was shown to Marshall in a place in the Exe valley in 1918 but has not been recorded there since. It was known in Leigh Woods (10) from 1789 till 1839 but the place was later enclosed. A number of other reports from North Somerset were probably errors for the next species.

Gymnocarpium robertianum (Hoffm.) Newm. Limestone Fern

Thelypteris robertiana (Hoffm.) Slosson; *Polypodium robertianum* Hoffm.; *Phegopteris calcarea* Fée

Rare and local on limestone rocks. It reaches the southern limit of its British distribution in a few places on Mendip (8 & 9). It still grows in Goblin Combe (9) but seems to have gone from a number of other places in North Somerset where it was recorded previously.

CYSTOPTERIS Bernh.

Cystopteris fragilis (L.) Bernh. Brittle Bladder-fern.

Fairly common on rocks and walls of the Carboniferous Limestone of Mendip but now only in a few other localities in the north of the county [Goblin Combe (9), Dundry Hill, Chew Magna, Stanton Drew, Lansdown and Bathford (10)] although formerly more widespread there. In the south and west only in a few places on walls, as at Simonsbath and near Bridgetown (1) and at Whitestaunton (6), where it may have been originally introduced. Not seen recently in districts 2, 3 or 4, and not recorded in 5 or 7.

ASPIDIACEAE

POLYSTICHUM Roth

Polystichum aculeatum (L.) Roth Hard Shield-fern

P. lobatum (Huds.) Chevall.

Fairly frequent in woods and on banks on Mendip and on the Jurassic limestones in the east and south of the county. Less common in the west. Absent from the lowlands and much of central Somerset.

× **setiferum** (*P.*× *bicknellii* (Christ) Hahne). Occurs where the parents grow close together but it is difficult to recognize and its distribution is still uncertain. Specimens from the south of the county have definitely been identified as this.

Polystichum setiferum (Forsk.) Woynar Soft Shield-fern

P. angulare (Willd.) C. Presl

Common in woods and on hedgebanks over much of the county, but rather scarce on Exmoor and in the lowlands.

DRYOPTERIS Adans.

Dryopteris filix-mas (L.) Schott Male Fern

Lastraea filix-mas (L.) C. Presl

Very common in woods and lanes, on hedgebanks and on walls.

× **pseudomas** (*D.* × *tavelii* Rothm.). This hybrid may occur where the parents grow together, but so far only one specimen from Somerset has been definitely identified.

Dryopteris pseudomas (Wollaston) Holub & Pouzar Scaly Male Fern

D. borreri auct.; *Lastraea filix-mas* var. *paleacea* T. Moore

Frequent in woods and on shady banks on Exmoor and the Brendon, Quantock and Blackdown Hills and on the Wiltshire border east of Bruton (8). Less common on the drier and lighter soils of North Somerset.

Dryopteris aemula (Ait.) Kuntze Hay-scented Buckler-fern

Lastraea aemula (Ait.) Brackenr.

Rare and only seen recently in wooded combes near Porlock and Holford (2). Old reports from the Axe valley near Winsham (6) have never been confirmed.

(**Dryopteris villarii** (Bellardi) Woynar (*Lastraea rigida* (Sw.) C. Presl) at one time grew sparingly on Hampton Down near Bath (10), but it had gone by 1866. It was probably planted there originally.)

†**Dryopteris cristata** (L.) A. Gray Crested Buckler-fern

Polypodium cristatum L.

Said by Sole in Collinson's *History* (1791) to grow in Emborough Wood (10) but not reported since.

Dryopteris carthusiana (Vill.) H. P. Fuchs Narrow Buckler-fern

D. lanceolatocristata (Hoffm.) Alston; *D. spinulosa* Watt; *Lastraea spinulosa* C. Presl

In wet woodland. Fairly frequent on the peat moors (8) but rare elsewhere and becoming more so. Not recorded recently from district 1.

Dryopteris austriaca (Jacq.) Woynar Broad Buckler-fern

D. dilatata (Hoffm.) A. Gray; *Lastraea dilatata* (Hoffm.) C. Presl

Common in woods, lanes and on shady banks everywhere except in the lowlands where it is only found on the peat moors (8).

× **carthusiana** (*D.* × *deweveri* (Jansen) Jansen & Wachter) occurs in damp, shady woodland, and as yet there are only a few records from Somerset.

BLECHNACEAE

BLECHNUM L.

Blechnum spicant (L.) Roth — Hard Fern

Lomaria spicant (L.) Desv.

Common on acid moors and in woodland on Exmoor, the Brendon, Quantock and Blackdown Hills and on the Greensand on the eastern border of the county. Less so on the Old Red Sandstone of Mendip and near Bristol. Very rare elsewhere. Not yet found in district 5.

MARSILEACEAE

PILULARIA L.

†**Pilularia globulifera** L. — Pillwort

Not seen in the county for many years. In 1916 Marshall was told it had been found near Winscombe (9) some years previously and he thought this might indicate that Sole's record in Collinson's *History* (1791) for "Blackdown" referred to Mendip, but the plant is also recorded for "Blackdown" in Polwhele's *History of Devonshire* (1797) which suggests the Blackdown Hills were intended. The only other record is for the Somerset Coal Canal at Monkton Combe (10), (H. F. Parsons, 1875). Few traces of this canal still exist.

AZOLLACEAE

AZOLLA Lam.

Azolla filiculoides Lam. — Water Fern

A tropical American plant which has found a congenial habitat in the rhines and ditches of the lowlands where it is now quite widespread. First reported in 1931 as having escaped from a pond on the Clevedon Court estate (9) where it had been introduced a few years previously. From there it spread to Kenn Moor and to Kewstoke by 1961. In 1939 it appeared on the Kennet & Avon canal just across the Wiltshire border and by the end of the year it had reached Bathampton (10). From Westonzoyland (5) in 1945 it has become established over much of Sedgemoor, and from near Brent Knoll (8 & 9) in 1946 it has spread to the coastal area and as far inland as Godney (8). Still scarce in South Somerset; it was first reported at Stolford (2) in 1951, then south-west of Burrow Bridge (3) in 1957, West Sedgemoor (3) in 1967 and Long Load (4) in 1971. It has also been at Minehead (2) since 1967.

SPERMATOPHYTA

GYMNOSPERMAE

PINACEAE

PINUS L.

Pinus sylvestris L. Scots Pine

Widespread throughout the county, usually planted but sometimes self-sown. Although pollen analysis of peat has shown that this species was once native in southern England it is highly unlikely that any descendants of the original forests survive in Somerset. Pines have been planted for at least 300 years and it is probable that trees thought by some to be possibly native are derived from planted ones.

(Other coniferous trees have been extensively planted, particularly on Exmoor, and on the Brendon, Quantock, Blackdown and Mendip Hills, by the Forestry Commission and other landowners. These include **Pseudotsuga menziesii** (Mirb.) Franco (Douglas Fir), **Picea abies** (L.) Karst. (Norway Spruce), **P. sitchensis** (Bong.) Carrière (Sitka Spruce), **Tsuga heterophylla** (Raf.) Sarg. (Western Hemlock), **Larix decidua** Mill. (European Larch), **L. kaempferi** (Lamb.) Carr. (Japanese Larch), **L.** × **eurolepis** Henry (Hybrid Larch), **Pinus nigra** Arnold ssp. **laricio** (Poir.) Maire (Corsican Pine), **Thuja plicata** D. Don (Western Red Cedar) and **Chamaecyparis lawsoniana** (A. Murr.) Parl. (Lawson's Cypress). Many other species are planted for ornament in parks and gardens. Single specimens of **Sequoiadendron giganteum** (Lindl.) Buchholz (Wellingtonia) are striking landmarks in several places.)

CUPRESSACEAE

JUNIPERUS L.

† **Juniperus communis** L. Juniper

Formerly abundant on the western slope of Bathford Hill above Warleigh (10) but the area has been cleared and replanted with conifers; it was last recorded there in 1946 (DEC). Another place where it was known for many years, between Hemington and Laverton (10), was ploughed up about 1965. Records near Uphill (9) (St. Brody, 1856) and above Street (8), 1921 (E. Ellman), were probably casual occurrences.

TAXACEAE

TAXUS L.

Taxus baccata L. Yew

Native on the Carboniferous Limestone and the Oolite in the north of the county where it is frequent on rocky outcrops and in woods. Widely planted, especially in churchyards, and often self-sown.

ANGIOSPERMAE

DICOTYLEDONES

RANUNCULACEAE

CALTHA L.

Caltha palustris L. Marsh Marigold

Marshes and by ponds and streams. Still quite common where suitable habitats exist.

HELLEBORUS L.

Helleborus foetidus L. Stinking Hellebore

A rare plant of calcareous woods, probably native on the Carboniferous Limestone near Cleeve and Churchill (9) and perhaps on the Oolite near Bath and Mells (10).
Introduced or a garden escape in other localities but well naturalized in some, as on the Polden Hills near Compton Dundon (5).

Helleborus viridis L. ssp. **occidentalis** (Reut.) Schiffn. Green Hellebore

A rare woodland plant, native in the north of the county in a few places but probably introduced or an escape elsewhere.

ERANTHIS Salisb.

Eranthis hyemalis (L.) Salisb. Winter Aconite

Naturalized in planted woodland in a few places.

RANUNCULACEAE

NIGELLA L.

Nigella damascena L. Love-in-a-mist

Garden escape.

ACONITUM L.

Aconitum napellus L. Monk's-hood

A. anglicum Stapf

By the Hillfarrance Brook (3), R. Yeo and tributaries (4), Alham and Brue (8) and Edford Brook and R. Frome (10). Quite often found in other places as an introduction or garden escape. Grown in gardens for centuries but it is believed to be native by streams in southwest England, and some authorities have considered our wild plant to be a separate endemic species. Another rather similar species, **A. variegatum** L. is also cultivated and has been found as an escape.

CONSOLIDA (DC.) S. F. Gray

Consolida ambigua (L.) P. W. Ball & Heywood Larkspur

Delphinium ambiguum L.; *D. ajacis* auct.

Grown in gardens and occasionally found as a casual.

ANEMONE L.

Anemone nemorosa L. Wood Anemone

Common in woods over most of the county but absent from much of Exmoor and rather rare in the lowlands.

Anemone apennina L. Blue Anemone

Occasionally naturalized in parkland.

(A note by Dillenius (circa 1740) says he had grown **Pulsatilla vulgaris** Mill. (*Anemone pulsatilla* L.) at Oxford from seeds picked between Bath and Bristol. White suggests it is not improbable that it once grew at the southern end of the Cotswolds above Kelston (10).)

CLEMATIS L.

Clematis vitalba L. Old Man's Beard, Traveller's Joy

An indicator of calcareous soils, so common over much of the county but absent from the acid soils of Exmoor and the Brendon, Quantock and Blackdown Hills and of Mendip near Priddy (9). Not seen recently in district 1.

RANUNCULUS L.

Ranunculus acris L. Meadow Buttercup

Abundant throughout the county in meadows, pastures and on roadside verges.

Ranunculus repens L. Creeping Buttercup

Abundant everywhere in fields, on roadsides and in waste places, and a weed in cultivated land and gardens.

Ranunculus bulbosus L. Bulbous Buttercup

Very common in meadows and pastures except on the higher ground of Exmoor.

Ranunculus arvensis L. Corn Buttercup

Formerly a widespread weed of cornfields but now very rare. The only recent records are near Somerton (5), 1970 (R & IR), Butleigh (8), 1965, and Keinton Mandeville (8), 1968 (CAH), Shoscombe (10), 1968 (R & IR), and as a casual on Steep Holm, 1965 (VG), at Bedminster (10), 1970 (AFD) and Brislington (10), 1979 (ALG).

Ranunculus sardous Crantz Hairy Buttercup

A rare weed of waste ground, particularly near the sea where it is more persistent than inland. Recent records are from Porlock Marsh (2), 1967 (RBG), Minehead (2), 1967 (SCH), Stolford (2), 1972 (R & IR), Yeovil Marsh (4), 1974 (CAH), Merriott (4), 1979 (R & IR), Aller (5), 1978 (CES), Brewham (8), 1976 (CAH) and Weston-super-Mare (9), 1976 (CES).

Ranunculus parviflorus L. Small-flowered Buttercup

A rare plant of dry slopes, mainly near the sea, as at Minehead, Blue Anchor and Stogursey in district 2, and Brean Down, Sand Point and Clevedon in district 9, but also on the steep Lias and Rhaetic hillsides near Fivehead (3), High Ham, Compton Dundon and Charlton Mackrell (5). Occasionally found elsewhere. It was particularly abundant in 1977 after the severe drought of the previous year had reduced competition from grasses and other plants in its usual habitats.

RANUNCULACEAE

Ranunculus auricomus L. Goldilocks

Quite frequent in woodland in the north and east of the county and on the Lias south of Taunton but rather rare elsewhere. Absent from most of Exmoor and not yet recorded from district 1.

Ranunculus lingua L. Greater Spearwort

A plant of stagnant fen water now very rare as a native but formerly more frequent on the Levels. It still survives in a moor ditch near Catcott (8) and in old ponds near Churchill and Kenn (9) but it has gone from the Gordano valley (9). Sometimes introduced in ponds; at Charterhouse on Mendip (9) it has flourished since 1921, and at Hinton Charterhouse (10) since 1973.

Ranunculus flammula L. Lesser Spearwort

Marshes and other wet places, preferring acid soils, so common on Exmoor and the Brendon, Quantock and Blackdown Hills, the Old Red Sandstone of Mendip, the Oxford Clay east of Bruton (7, 8 & 10) and on the peat of the Levels. Occasional on the Coal Measures. Rare elsewhere.

Ranunculus sceleratus L. Celery-leaved Buttercup

Common throughout the lowlands by ditches and streams and at the edges of ponds. Rare on higher ground and absent from districts 1 and 6.

Ranunculus hederaceus L. Ivy-leaved Crowfoot

On muddy tracks and by small streams and pools on acid soils. Fairly frequent in the west and south, on the Mendip sandstone and on the Greensand of the eastern border but very rare elsewhere. A decreasing species. Although White describes it as "Rather Common" it has gone from most of his North Somerset localities.

Ranunculus omiophyllus Ten. Round-leaved Crowfoot

R. lenormandii F. W. Schultz

Rather common in marshy places and by pools on the moorlands of Exmoor and the Brendon, Quantock and Blackdown Hills, with an outlying locality near Chard (6). Absent from the rest of Somerset. The dot shown in the *B.S.B.I. Atlas* (1962) for 10 Km. square 31/63 is an error.

[**Ranunculus fluitans** Lam. has only once been reliably reported from the county, a casual occurrence in the Avon near Newton St. Loe (10), 1917 (IMR).]

Ranunculus circinatus Sibth. Fan-leaved Water-crowfoot

Fairly common in moor ditches and ponds in the Levels between Kingsbury Episcopi (4) and Uphill (9) and less so in the north round Yatton and in the Gordano valley (9). Very rarely in running water. The only recent records away from the main area of distribution are in a pond near Hardington, 1966 (EBH), and in the Sutton Bingham reservoir, 1971 (JGK), in district 4 and in the Kennet and Avon canal (10) where it has been known for a hundred years.

Ranunculus trichophyllus Chaix Thread-leaved Water-crowfoot

R. drouetii F. W. Schultz ex Godr.

Fairly frequent in ditches and pools throughout the lowlands and occasionally elsewhere. Not recorded in districts 1 or 6 and not seen recently in district 2 west of Stolford. This and the next two species are characteristic pioneers of newly-cleaned or recently-dug ditches and ponds.

Ranunculus aquatilis L. Common Water-crowfoot

R. heterophyllus Weber; *R. radians* Revel

In similar, and sometimes the same, places as the last species but rather more widely distributed. Not recorded in districts 1 or 6.

Ranunculus peltatus Schrank

R. aquatilis ssp. *peltatus* (Schrank) Syme; *R. floribundus* Bab.

Ditches and ponds; rather rare and apparently decreasing, but its present distribution is uncertain because of confusion with the last species. Records for this in rivers are probably forms of the next species.

Ranunculus penicillatus (Dumort.) Bab.

R. pseudofluitans (Syme) Newbould ex Bak. & Foggitt

In rivers and streams in all districts. Very variable and has been regarded as a subspecies of either *R. aquatilis* or *R. peltatus*. The most frequent form in Somerset rivers, where the water is usually base-rich, is var. **calcareus** (R. W. Butler) Cook, which has no floating leaves. Var.

penicillatus, with floating leaves, occurs in the Barle in district 1 and may be found in other rivers where the water is base-poor.

Ranunculus baudotii Godr. Brackish-water Crowfoot

R. confusus Godr.

Quite frequent in pools and ditches near the sea, usually in brackish water.

Ranunculus ficaria L. Lesser Celandine

Very common on hedgebanks and roadsides and in woods and damp places. Ssp. **bulbifera** Lawalrée seems equally as common as ssp. **ficaria** and attempts to distinguish habitat preferences have been unsuccessful.

[**Adonis annua** L., never more than a rare casual in Somerset, was last seen in 1938 near Bath (10) (AEW)].

Myosurus L.

Myosurus minimus L. Mousetail

Only known as a very rare casual in Somerset until 1920 when it was found (WW) on a drove on West Sedge Moor (3). It has persisted there and is now quite widespread on the alluvial clay both sides of the moor (not on the peat in the centre), abundant in some years. It is apparently spread by tractor tyres. In 1974 it was found at Ash (4) (JGK) on a muddy roadside and was still there in 1980. The Water Authority had been carrying out drainage work nearby and it may have been brought in from West Sedge Moor by some of the earth-moving machines.

Aquilegia L.

Aquilegia vulgaris L. Columbine

Probably native only in woods on the limestones in the north and east of the county, but widespread and quite frequent as a garden escape.

Thalictrum L.

Thalictrum flavum L. Common Meadow-rue

Abundant on West Sedge Moor (3) and fairly frequent in marshy places and by streams in the basins of the Parrett and Brue and in the Cheddar valley. Rare elsewhere in marshes and on river banks.

Thalictrum minus L. ssp. **minus** Lesser Meadow-rue

Plentiful in the Cheddar Gorge (9). A garden escape elsewhere.

PAEONIACEAE

PAEONIA L.

Paeonia mascula (L.) Mill. Peony

P. corallina Retz

Established on Steep Holm where it was first reported in 1803 by F. B. Wright. There were then some buildings on the island used by fishermen but there is no record of permanent habitation after the ancient Priory was abandoned about the end of the 13th century until 1832, so the date of its introduction is unknown. It is not mentioned by Banks in his account of a visit to the island with Lightfoot on 3rd July, 1773, but it was almost certainly already there then. The plant grows on the steep eastern slopes where there is some protection from the westerly gales in winter and where the only springs on the island exist. At one time it was quite plentiful but by 1891 Murray estimated there were less than 20 plants. Of late years its territory has been invaded by sycamores and by 1978 only one plant survived. Some of the trees have been cleared and seeds from the plant sown in an attempt to re-establish the species in this its only British locality as a naturalized alien. At one time it was cultivated, the roots being used in medicine, and Babington (1839) records an escape near Bath just over the Gloucestershire border.

BERBERIDACEAE

EPIMEDIUM L.

Epimedium alpinum L. Barren-wort

An occasional garden escape. At one time established in Leigh Woods (10) where it was first reported by Dr H. O. Stephens. A specimen collected there about 1850 by Miss I. Gifford is in **TTN.**

BERBERIS L.

Berberis vulgaris L. Barberry

Widely planted in hedges and occasionally birdsown in disused quarries, etc. Much less frequent since it has been proved to be the host of a rust harmful to wheat. Not seen recently in districts 3 or 7 and not recorded in districts 1 or 6.

Berberis glaucocarpa Stapf

B. aristata auct.

Much planted for hedging round Porlock and Minehead (2) and occasionally birdsown in that area.

MAHONIA Nutt.

Mahonia aquifolium (Pursh) Nutt. Oregon Grape

Grown in gardens and sometimes planted in woodland as game cover. Naturalized escapes from these sources are occasionally found.

NYMPHAEACEAE

NYMPHAEA L.

Nymphaea alba L. White Water-lily

Often planted and long persistent in ornamental ponds and lakes and sometimes in more remote pools where deliberate introduction is less obvious. Reported in the canal basins at Bath by Dr H. Gibbes (Babington, 1834) it now extends throughout the Somerset stretch of the Kennet & Avon canal (10). The only recent occurrence in the county in a river is in the R. Isle near Curry Rivel and Drayton (4). There are no recent records from districts 1, 2 or 3.

Other cultivated water-lilies are occasionally found as escapes. One with pink flowers is well established in the R. Tone at Creech St Michael (3).

NUPHAR Sm.

Nuphar lutea (L.) Sm. Yellow Water-lily

Nymphaea lutea L.

Frequent in the rivers of central and south Somerset and in the Avon and its tributaries but absent from the smaller rivers of districts 2 and 9. Also in the Bridgwater & Taunton (3) and Kennet & Avon (10) canals. Sometimes planted in ponds. Not recorded in districts 1, 2 or 9 except as an introduction.

CERATOPHYLLACEAE

Ceratophyllum L.

Ceratophyllum demersum L. Rigid Hornwort

Locally common in moor ditches and rhines of the Levels from Kingsbury (4), Long Sutton (5) and Glastonbury (8) to the coast, and from Nailsea (9) to Weston-super-Mare and Clevedon. Rare in similar places in the Cheddar and Gordano valleys (9). Plentiful in the Bridgwater & Taunton (3) and Kennet & Avon (10) canals. Formerly recorded in the Rivers Frome and Avon (10) but not seen recently in either. It occurs occasionally in ponds, an introduction in some at least. The only records in districts 6 and 7 are from ponds [Bishopswood (6), 1971 (NGC) and Gasper (7), 1967 (CAH)] and there are no records from districts 1 or 2.

Ceratophyllum submersum L. Soft Hornwort

Rather rare, in ditches and rhines mostly near the coast, from Wall Common (2) and Wembdon (3) to Lympsham (9) and from Kewstoke to Clevedon (9). There are isolated records further inland at Kingsbury, 1957 (EJH) and Drayton, 1970 (JGK) in district 4, at Glastonbury (8), 1970 (JA), where it was first noted by Dillenius in 1726, and from the Kennet & Avon canal (10), 1931 (WDM).

PAPAVERACEAE

Papaver L.

Papaver rhoeas L. Common Poppy

Although much reduced in cornfields by the use of weedkillers it still occurs in them. Very common in waste places and by roadsides throughout most of the county except for the higher parts of Exmoor. Never yet recorded in district 1. The form with adpressed hairs (forma **strigosum** Boenn.) occurs quite frequently.

Papaver dubium L. Long-headed Poppy

Frequent and widespread but less common than *P. rhoeas.* Rare west of Taunton, the only recent records being at Bossington (2), 1970 (R & IR), Treborough (2), 1973 (CAH) and Wellington (3), 1978 (CAH). Not yet recorded in district 1.

PAPAVERACEAE

Papaver lecoqii Lamotte — Babington's Poppy

Fairly frequent in waste places on the Lias and Oolite of central and east Somerset. Rare or overlooked elsewhere. Not recorded recently in district 9 and never in 1 or 6.

†**Papaver hybridum** L — Rough Poppy

At one time abundant in cornfields round Bridgwater (2, 3 & 5) and Burnham (8) it had become very rare by the end of the 19th century and must now be presumed extinct in the county. The last record is for one plant in a cornfield at Charlton Mackrell (5) in 1948 (FKM).

Papaver argemone L. — Pale Poppy

Formerly a cornfield weed but now only a rare casual in waste places. The only recent records are from Compton Dundon (5), 1979 (RGR), Edington, 1970 (HWB), Puriton, 1971 (TF) and Highbridge, 1975 (RSC) in district 8, Berrow (9), 1962 (RB) and Hinton Charterhouse (10), 1968 (RBG). There are old records for districts 2, 3 and 4 but it has not been seen in V.C.5 since 1936 when it was recorded at Roadwater (2) (EJH and WW).

Papaver somniferum L. — Opium Poppy

A frequent casual of waste places and rubbish tips and a weed in gardens. Recorded in all districts but less frequent in the west than in the rest of the county.

Papaver lateritium C. Koch — Armenian Poppy

An occasional garden escape.

MECONOPSIS Vig.

Meconopsis cambrica (L.) Vig. — Welsh Poppy

Native in rocky and shady Exmoor valleys, particularly those of the Exe and Barle in district 1, and in the Cheddar Gorge (9). A garden escape elsewhere.

GLAUCIUM Mill.

Glaucium flavum Crantz — Yellow Horned Poppy

On sand and shingle at intervals along the coast of district 2 from Bossington to Steart. It was already decreasing on the North Somerset coast in 1912 according to White but has now completely gone. It was last recorded at Burnham (8) in 1924 (HST) and at Kewstoke Bay (9) in 1917 (Miss Livett).

CHELIDONIUM L.

Chelidonium majus L. Greater Celandine

Common on hedgebanks, walls and roadsides, usually near houses.

ESCHSCHOLZIA Cham.

Eschscholzia californica Cham. Californian Poppy

Garden escape.

FUMARIACEAE

CORYDALIS Vent.

Corydalis claviculata (L.) DC. Climbing Corydalis

On acid stony and sandy soils. Quite frequent on the borders of Exmoor and the Brendons, at the southern end of the Quantocks and on the Greensand between Penselwood (7) and Brewham (10). It was said by White to have disappeared in the 1890s from an isolated locality on the Coal Measures near Temple Cloud (10) where it had been known since 1839 but it was refound in 1972 and 1973 in small quantity in two spots near by (PA & MH).

Corydalis lutea (L.) DC. Yellow Corydalis

Introduced a very long time ago but now commonly naturalized on walls in towns and villages throughout the county.

FUMARIA L.

Fumaria capreolata L. White Ramping Fumitory

F. pallidiflora Jord.

Usually on hedgebanks but sometimes on cultivated or waste ground. Common near the coast of district 2 between Bossington and Dunster, and here and there as far east as Kilve. Long known near Axbridge and Locking in district 9, and it persisted on Steep Holm from 1891 till at least 1910 but has now gone. A rare casual elsewhere; the only recent records are at Sparkford (5), 1967 (CAH), Butleigh (8), 1967 (CAH) and Portbury (9), 1974 (R & IR).

FUMARIACEAE

†**Fumaria purpurea** Pugsl. Purple Ramping Fumitory

Extinct in Somerset. The only records are as a weed in cultivated ground near Huish Champflower (3),1949 (FH), and on the hillside at Christon (9), before 1912 (Mrs E. S. Gregory).

Fumaria bastardii Bor. Tall Ramping Fumitory

Hedgebanks and cultivated ground. Rare and local. Only known in the county about Minehead (2), near Cheddar (9), 1930 (C & NS), and at Compton Dando (10), 1976 (E. Paskin conf. M. G. Daker). This is the species called *"F. muralis"* in Murray (1896).

†**Fumaria martinii** Clavaud

This very rare colonist of arable land has only once been found in Somerset, at Combe Down, 1920 (CIS). It has not been seen since.

Fumaria muralis Sond. ssp. **boraei** (Jord.) Pugsl.
Common Ramping Fumitory

F. confusa sensu Murray & White

A common weed of arable land and waste places in districts 2 and 3 but very rare elsewhere. The only recent records from other districts are near Upton (1), 1975 (HWB), Marston Magna (5), 1966 (CAH), and near Shapwick (8), 1970 (CAH) and 1978 (RGR).

Fumaria densiflora DC. Dense-flowered Fumitory

F. micrantha Lag.

A very rare casual in Somerset. Recorded only at Taunton (3), 1974 (CMA), and at Bedminster (10), 1922 (C & NS).

Fumaria officinalis L. Common Fumitory

A common weed of arable land throughout the county except on Exmoor. Not seen recently in district 1. The ssp. **wirtgenii** (Koch) Arcang. has only been detected twice, at Knowle (10), 1881 (W. H. Painter in **BM**), and at Keynsham (10), 1918 (IMR).

CRUCIFERAE

Brassica L.

†**Brassica oleracea** L. Wild Cabbage

Although recorded in district 9 from Berrow, Brean and Steep Holm by J. C. Collins (1837) it was never seen in Somerset by Murray nor White. T. B. Flower recorded it on Steep Holm in 1887 but it was not reliably reported there on later visits of which there have been many by well-known botanists. It is doubtful whether it ever occurred in the county as a truly wild plant but if so it is certainly extinct now.

Brassica napus L. Rape

Frequently grown as a fodder crop so widespread as an escape from cultivation. Locally common and more or less naturalized on the banks of the Parrett, Yeo and Brue in central Somerset in districts 4, 5 and 8.

Brassica rapa L. Wild Turnip

B. campestris L.

Common as an escape from cultivation over much of the county. Long established on some river banks, particularly those of the Avon and Chew in the north (10), and it may possibly be native there. It makes a fine show in early summer on the cliffs of Steep Holm where it was first collected by T. Clark about 1832.

Brassica juncea (L.) Czern. Chinese Mustard

An occasional casual, first recorded in the county in 1922 at Orchard Wyndham (2) (Miss A. Wyndham).

Brassica nigra (L.) Koch Black Mustard

Common near the coast, on river banks and throughout the lowlands on field borders and in waste places. Recorded in all districts but not seen recently in district 1.

Erucastrum C. Presl

Erucastrum gallicum (Willd.) O. E. Schultz Hairy Rocket

Brassica gallica (Willd.) Druce

A rare casual, first recorded in the county in 1916 at Long Ashton (10)

(IMR). The only recent records, both in district 10, were in arable land at Mells, 1969 (RGR), and on a railway bank near Bristol, 1971 (AFD).

Sinapis L.

Sinapis arvensis L. Charlock

A common weed of arable land and waste ground, less frequent on Exmoor than elsewhere in the county.

Sinapis alba L. White Mustard

Waste places and rubbish tips. Uncommon but recorded in all districts except 1. Now only rarely found in the county as a weed of arable land.

Hirschfeldia Moench

Hirschfeldia incana (L.) Lagr.-Foss. Hoary Mustard

Sinapis incana L.

A rare casual, first reported in the county in 1897 near Twerton (10) by S. T. Dunn. It has become well established on waste ground at Weston-super-Mare (9) since it was first seen there in 1962 (RMB). Other records are at Cannington (3), 1958 (CMA), Charlton Mackrell (5), 1978 (RGR), Berrow (9), 1951 (C & NS), and Brislington (10), 1979 (ALG).

Diplotaxis DC.

Diplotaxis muralis (L.) DC. Annual Wall-rocket

A weed of waste places, very common near the sea from Highbridge (8) to Bristol (10). Widespread, particularly near railways, but less common elsewhere. White (1912) remarked on its increase in the west of England. It has certainly spread since Murray's Flora was written (1896) to districts 2, 5, 6 and 7 but has not yet been found in district 1.

Diplotaxis tenuifolia (L.) DC. Perennial Wall-rocket

On old walls and in waste places. Common in the north around Bath and Bristol (10). Rare elsewhere and usually by railway tracks, as at Wellington, Taunton (3) and Charlton Mackrell (5). Not in districts 1, 4, 6 or 7, and not seen since 1931 in district 2.

RAPHANUS L.

Raphanus raphanistrum L. Wild Radish

A widespread but not common weed of arable land, roadsides and waste places. In cultivated land it is usually white-flowered, while on roadsides the yellow-flowered form is locally common. Both varieties may be found in waste places.

Raphanus maritimus Sm. Sea Radish

Formerly very rare but now quite frequent near the Channel coast from Steart (2) and Combwich (3) to Portishead (9).

CRAMBE L.

†**Crambe maritima** L. Sea Kale

Collected at Burnham (8) about 1835 by T. Clark and seen there at a later date by T. B. Flower but now long extinct. Recorded at Brean Down by St Brody (1856). A plant found on a tip at Corston (10), 1950 (IWE), was probably a garden outcast.

ERUCA Mill.

Eruca vesicaria (L.) Cav. ssp. **sativa** (Mill.) Thell. Garden Rocket

E. sativa Mill.

A rare casual, first recorded in the county at Twerton (10), 1897 (S. T. Dunn), and only seen recently on a tip at Brislington (10), 1978 (ALG).

RAPISTRUM Crantz

Rapistrum rugosum (L.) All. Bastard Cabbage

Including *R. orientale* (L.) Crantz and *R. hispanicum* (L.) Crantz

A rare casual. Recent records are in districts 4, 5, 9 and 10 only.

CAKILE Mill.

Cakile maritima Scop. Sea Rocket

Formerly common on sandy and shingly shores at about the High Water mark but now rather scarce and only between Steart (2) and Kewstoke (9). White remarked on its decrease under human pressure and the process has since gone much further.

CRUCIFERAE

LEPIDIUM L.

Lepidium sativum L. Garden Cress

A frequent casual on rubbish tips.

Lepidium campestre (L.) R.Br. Field Pepperwort

Arable fields and waste places. Locally frequent in the central lowlands of districts 3, 4 and 5 and in the Gordano (9) and Avon (10) valleys of the north. Rare elsewhere. Not recorded recently in district 6 and never in district 1.

Lepidium heterophyllum Benth. Smith's Pepperwort

L. smithii Hook.

On roadsides and hedgebanks on dry acid soils. Fairly frequent on Exmoor. It has not been reliably reported except in districts 1 and 2 for more than 50 years although it was formerly recorded in districts 3, 4, 9 and 10 also.

Lepidium ruderale L. Narrow-leaved Pepperwort

A rare denizen of docks and waste places near the sea. Still to be found at Bridgwater (3), 1966 (HWB), Stert Island (2), 1974 (RSC), Weston-super-Mare (9), 1962 (RMB), Portishead (9), 1974 (CHC), and Ashton Gate (10), 1969 (AFD). An occasional casual inland. Recent records are at Muchelney (4), 1968 (CAH), and Bruton (8), 1963 (CAH).

†**Lepidium latifolium** L. Dittander

A salt-marsh plant which was known at or near Berrow (9) since 1835. It became established on a sandy hedgebank near the sea by at least 1860 when specimens were collected by T. Clark and T. B. Flower, and survived there for a century. It was seen there in 1960 but the spot has since been overwhelmed by building.

CORONOPUS Haller.

Coronopus squamatus (Forsk.) Aschers. Swine Cress

Senebiera coronopus (L.) Poir.

A common weed near farm gateways and other well-trampled places throughout the county except for the higher parts of Exmoor and the Brendon, Quantock and Mendip Hills. Not recorded in district 1.

Coronopus didymus (L.) Sm. Lesser Swine Cress

Senebiera didyma (L.) Pers.; *S. pinnatifida* DC.

A common weed of waste places which appears to have spread in recent years. Recorded in all districts. Now abundant on Steep Holm but not known there before 1964.

CARDARIA Desv.

Cardaria draba (L.) Desv. Hoary Cress

Lepidium draba L.

This alien weed of waste places was first recorded in Somerset near Keynsham (10) in 1886 although it was collected in Bristol (Glos.) in 1846. It has since become widespread, particularly near railways, and is now recorded in all districts except 6. Occasionally in arable land where it is very difficult to eradicate.

ISATIS L.

†**Isatis tinctoria** L. Woad

Still cultivated about Keynsham (10) in 1791 according to Sole but growing it there must have ceased early in the 19th century. T. B. Flower told White he had found a plant or two once or twice in the district but it has not been seen since in the county except once as a casual at Weston-super-Mare (10), 1906 (Rev S. Laing).

IBERIS L.

Iberis amara L. Candytuft

Only a garden escape in Somerset, as is **I. umbellata** L.

THLASPI L.

Thlaspi arvense L. Field Penny-cress

A widespread weed of arable land, apparently more frequent than it was formerly. Recorded from all districts but rare on the higher ground.

Thlaspi perfoliatum L. Perfoliate Penny-cress

Established on railway banks at Charlton Mackrell (5) where it was first noticed in 1965 (CAH). It has maintained its ground but not spread from the places in which it was originally found.

CRUCIFERAE

Thlaspi alpestre L. Alpine Penny-cress

Locally plentiful on the Mendip limestone between Charterhouse and Sandford (9), particularly on old mining ground.

TEESDALIA R.Br.

†**Teesdalia nudicaulis** (L.) R.Br. Shepherd's Cress

A plant of sandy or gravelly places only twice recorded in Somerset and now apparently extinct. The older locality, near East Chinnock (4), prior to 1896 (Rev J. Sowerby), was long lost, rediscovered in 1924 (H. Downes & WDM) but it has not been seen there since. It was found on Porlock Hill (2) in 1916 (NGH) and last reported there in 1929. An 1856 record from Cheddar (9) was almost certainly an error.

CAPSELLA Medic.

Capsella bursa-pastoris (L.) Medic. Shepherd's Purse

An abundant weed of cultivated ground and roadsides, found almost everywhere.

HORNUNGIA Reichb.

Hornungia petraea (L.) Reichb. Hutchinsia

Hutchinsia petraea (L.) R.Br.; *Lepidium petraeum* Sm.

A very rare plant of limestone rocks now only to be found at Charterhouse (9) and in the Avon Gorge (10). Sole (1791) records it at Worle and Uphill (9) in both of which places it may have been destroyed by quarrying, and Collins (1837) at Cheddar also. The latter may refer to the Charterhouse locality not far away where it was first reported in 1935 (W. S. Parry). The occurrence at Millard's Hill (10) in a garden where there are many introduced plants [G. B. Milne-Redhead in Marshall (1914)] was probably accidental; it has since died out there.

COCHLEARIA L.

Cochlearia officinalis L. Common Scurvy-grass

Locally common on and under limestone rocks along the coast from Brean Down to Portishead, on the cliffs of Steep Holm and in the Cheddar Gorge (9). On walls at Shepton Mallet (8) and a very occasional casual elsewhere inland.

Cochlearia danica L. Danish Scurvy-grass

In rocky places and on banks near the sea. Bossington, Minehead, Dunster and Watchet (2), and from Brean Down and Weston-super-Mare, where it is quite plentiful, to Sand Point (9). One of the first species to be recorded on Steep Holm, about 1688 by James Newton (Ray, 1724) but it has not been seen there since 1963 (VG).

Cochlearia anglica L. English Scurvy-grass

Locally common on the muddy banks of the estuaries of the Parrett (3), Brue (8), Axe, Yeo (9), and Avon (10), and occasionally in salt marshes, as at Berrow and Sand Bay (9).

BUNIAS L.

Bunias orientalis L. Warty Cabbage

A rare casual, occasionally seen about Bristol. The only recent record was in Ashton Park (10), 1974 (RGR).

AUBRIETA Adans.

Aubrieta deltoidea (L.) DC. Purple Rock-cress

Garden escape.

LUNARIA L.

Lunaria annua L. Honesty

A frequent and widespread garden escape.

LOBULARIA Desv.

Lobularia maritima (L.) Desv. Sweet Alison

Alyssum maritimum (L.) Lam.

A frequent garden escape, especially persistent near the sea, as about Minehead (2) and Weston-super-Mare (9).

DRABA L.

Draba muralis L. Wall Whitlow-grass

A very rare plant of rocky limestone outcrops at the eastern end of Mendip near Pilton (8), Charterhouse (9) and Emborough (9), sometimes naturalized and very persistent on old walls in the same area but less often now than formerly. It was also known on the chalk near Chard (4 & 6) but has not been seen there since 1926, and it was once naturalized on a wall near Yeovil (4) but has long since disappeared. A weed in the Botanic Gardens and at Prior Park, Bath (10).

Erophila DC.

Erophila verna (L.) Chevall. Whitlow-grass

E. vulgaris DC.; *Draba verna* L.

Wall-tops, disused railway tracks and other dry places. Common over much of the county, particularly on limestone, and recorded in all districts but scarce over much of the Levels and very rare on Exmoor. The ssp. **spathulata** (Lang) Walters (*Draba brachycarpa* Jord.) occurs on the sand dunes at Minehead (2) and Burnham (8) and probably elsewhere on limestone but its distribution is uncertain.

Armoracia Gilib.

Armoracia rusticana Gaertn., Mey. & Scherb. Horse-radish

A common escape from cultivation, recorded in all districts.

Cardamine L.

Cardamine pratensis L. Cuckoo Flower

Very common in all districts in damp meadows, marshes and on roadside verges.

Cardamine impatiens L. Narrow-leaved Bitter-cress

Very rare and sporadic in appearance in shady places on the limestones. Only recorded recently in rides in Asham and Friary Woods (10). It has a long history in the Cheddar Gorge (9) but has not been seen there since 1950. In six or seven other localities scattered about in districts 9 and 10 it did not seem to persist.

Cardamine flexuosa With. Wavy Bitter-cress

Very common in damp shady places and by streams over most of the county but less frequent in the Levels in districts 3, 4 and 5.

× **pratensis** (*C.* × *haussknectiana* O. E. Schulz). Although the parent species sometimes grow together hybrids are apparently rare but they have occurred in V.C.5.

Cardamine hirsuta L. Hairy Bitter-cress

An abundant weed of gardens, cultivated land and waste places, and on wall tops.

Cardamine bulbifera (L.) Crantz Coralroot

Dentaria bulbifera L.

On hedgebanks at Trudoxhill and in a wood at Bath (10), in both places escaped from the grounds of large houses near by. The Bath locality is not far from Prior Park where it was recorded by Sole but disappeared early in the 19th century.

BARBAREA R.Br.

Barbarea vulgaris R.Br. Winter-cress

Common by rivers, streams and roadside ditches, except on the higher ground of Exmoor and the Brendon, Blackdown, Quantock and Mendip Hills. Not recorded recently in district 1.

Barbarea intermedia Bor. Medium-flowered Winter-cress

A rare weed of arable land and waste places less frequent than it was formerly. The only recent records are from a building site at Yeovil (4), 1968 (DS), a cornfield at Compton Dundon (5), 1977 (RGR), a rubbish tip at Dulcote (8), 1973 (CAH), and a field gateway near Stowey (10), 1968 (RGR). There are earlier records from districts 2, 3, 6 and 9.

Barbarea verna (Mill.) Aschers. American Winter-cress

B. praecox (Sm.) R.Br.

A rare escape from cultivation, formerly more widespread, but the only recent records are from near Minehead (2), 1974 (CJG), and Taunton (3), 1978 (RGR).

ARABIS L.

Arabis caucasica Schlecht.

A garden plant often established on walls and sometimes naturalized on hedgebanks.

Arabis hirsuta (L.) Scop. Hairy Rock-cress

On rock outcrops, in quarries and on old walls in the limestone areas.

Common on Mendip and quite frequent elsewhere in the north of the county. Rare in the south. It has not been seen recently on the Lias near Watchet (2), on the Polden Hills (5 & 8) nor on the chalk near Chard (4 & 6) although previously recorded for all of these. Not yet found in districts 1 or 3.

Arabis stricta Huds. Bristol Rock-cress

A. scabra All.

Only on the limestone of the Avon Gorge (10). Old records for Bath (Johnson, 1634) and Cheddar (Syme, 1863-92) are believed to be errors for *A. hirsuta*. The plant was introduced about 1920 by H. Corder on old walls near Bridgwater and Cannington (3) where it still survives.

This species and several others of the Somerset limestone rarities were the subjects of a transplant experiment started in 1955 by the Botany Department of Bristol University. The localities chosen were kept secret and the results of the experiment have not yet been published.

†**Arabis glabra** (L.) Bernh. Tower Mustard

A. perfoliata Lam.; *Turritis glabra* L.

Recorded by the Rev G. Swayne from old quarries above Bath (Withering, 1796) but not reported since.

NASTURTIUM R.Br.

Nasturtium officinale R.Br. Water-cress

Rorippa nasturtium-aquaticum (L.) Hayek; *Radicula officinalis* Groves

Very common by streams and ditches in all districts.

Nasturtium microphyllum (Boenn.) Reichb.

N. uniseriatum Howard & Manton; *Rorippa microphylla* (Boenn.) Hyland

This species has only been recognized by British botanists since 1946 and its distribution is still imperfectly known as it is overlooked by some recorders. It appears to be quite frequent in central and north Somerset, and has so far been identified in districts 4, 5, 8, 9 and 10.

× **officinale** (*Rorippa* × *sterilis* Airy Shaw). The hybrid can be recognized by the deformed and dwarfed fruit, few of which have more than one good seed. Sometimes grown commercially, it spreads vegetatively so is often found in the absence of one or both parents. There are scattered records from all districts except 2, 6 and 7.

RORIPPA Scop.

Rorippa sylvestris (L.) Bess. Creeping Yellow-cress

Nasturtium sylvestre (L.) R.Br.; *Radicula pinnata* (L.) Moench

Banks of the Tone, Parrett (3) and Huntspill (8) rivers and on the shores of the Blagdon (9) and Chew Valley (10) reservoirs. Occasionally on or by droves in the Levels of districts 8 and 9. A weed in gardens, nurseries and waste places in a few scattered localities in districts 2, 5, 6 and 9, and particularly in and around Bath (10) where it was formerly reported by the Avon and is still frequent by the By Brook, a small tributary.

Rorippa palustris (L.) Bess. Marsh Yellow-cress

R. islandica auct.; *Nasturtium palustre* (L.) DC.; *Radicula palustris* (L.) Moench

On muddy waste ground and by rivers streams and ponds. Not common except on the peat moors (8). Very rare west of Taunton (3) and absent from districts 1 and 2.

Rorippa amphibia (L.) Bess. Great Yellow-cress

Nasturtium amphibium (L.) R.Br.; *Armoracia amphibia* (L.) G. F. W. May

On and near the banks of the Avon from the Wiltshire border to Brislington (10). Occasionally by ponds and ditches in other parts of the county, the only recent such record being at Aller (5), 1972 (CAH). Recorded in the past by other rivers including the Tone (3), Isle (4), Parrett (5) and Brue (8) but it has not proved permanent by any of them. The most recent example of this is the King's Sedgemoor Drain (the artificially-modified lower part of the River Cary) in district 5, where it was seen in abundance about 1959 but has since disappeared (BS).

× **sylvestris** (*R.* × *anceps* (Wahlenb.) Reichb.). This hybrid was collected at Catcott (8) in 1920 by H. F. Devis (**K**) soon after *R. amphibia* was first seen on the peat moor (1919, CIS). It is very likely to occur by the Avon near Bath (10) but has not been identified there yet.

CRUCIFERAE

MALCOLMIA R.Br.

Malcolmia maritima (L.) R.Br. Virginian Stock

Garden escape.

HESPERIS L.

Hesperis matronalis L. Dame's Violet

A widespread garden escape usually seen by roadside ditches. Not recorded recently in district 6.

ERYSIMUM L.

Erysimum cheiranthoides L. Treacle Mustard

A weed of arable land and waste places, quite frequent on the peat moors (8) and in other parts of the lowlands of districts 4, 5, 8 and 9. A rare casual in other districts, particularly in the west. Not yet recorded in districts 1 or 7.

CHEIRANTHUS L.

Cheiranthus cheiri L. Wallflower

A common garden escape, naturalized on walls and often of long standing as at Glastonbury Abbey (8) where it was collected by T. Clark in 1829. Sometimes on rocks and old quarry faces. Conspicuous on the more sheltered cliffs of Steep Holm.

ALLIARIA Scop.

Alliaria petiolata (Bieb.) Cavara & Grande Garlic Mustard

A. officinalis Andrz. ex Bieb.; *Sisymbrium alliaria* (L.) Scop.

Abundant on hedgebanks and by roadsides throughout the county.

SISYMBRIUM L.

Sisymbrium officinale (L.) Scop. Hedge Mustard

Very common by roadsides and in waste places in all districts except on Exmoor.

Sisymbrium irio L. London Rocket

A rare casual, well established in and around a car park at Taunton (3),

1976 (RGR). Credited to Somerset on the authority of Gapper in the *New Botanist's Guide* (1835) but the only previous localized records were at Portishead Dock (9), 1909 (IMR in **LDS**), and Bedminster (10), 1922 and 1930 (CIS).

Sisymbrium orientale L. Eastern Rocket

S. columnae Jacq.

An alien, first recorded at Portishead (9) in 1896 (JWW), which apparently began to spread in waste places about 1924 when it was reported at Taunton (3), Highbridge (8) and Weston-super-Mare (9). Now recorded from all districts except 1 and 7. It can hardly claim to be firmly established yet except perhaps about Minehead (2) and in Bristol (10), especially between the old and new courses of the Avon (V.C.34).

Sisymbrium altissimum L. Tall Rocket

S. pannonicum Jacq.

A rare casual only reported recently at Camerton (10), 1965 (RGR), and at St Anne's, Bristol (10), 1971 (AFD). It was more widespread in the 1920s when there were eight records in various parts of the county. It was common about Bristol in 1932 (CIS) but the only other record since then was at Hutton (9), 1955 (NYS).

ARABIDOPSIS (DC.) Heynh.

Arabidopsis thaliana (L.) Heynh. Thale Cress

Sisymbrium thalianum (L.) Gay

A common weed in gardens and other cultivated land and on wall tops in many parts of the county, particularly on sandy and light soils. Rare on the alluvial clay of the Levels.

CAMELINA Crantz

Camelina sativa (L.) Crantz Gold of Pleasure

Formerly quite a frequent casual but now only seen as a bird-seed alien on rubbish tips.

CRUCIFERAE

DESCURAINIA Webb & Berth.

Descurainia sophia (L.) Webb ex Prantl Flixweed

Sisymbrium sophia L.

An alien long established in sandy waste places between Highbridge (8) and Berrow (9) but very rare there now. A casual elsewhere. The only other recent records are from Dunster (2), 1967 (SCH), Taunton (3), 1970 (CMA), and Frome (10), 1979 (AD).

RESEDACEAE

RESEDA L.

Reseda luteola L. Weld

A common weed of waste places about the larger towns and also in limestone quarries. Recorded in all districts except 1.

Reseda lutea L. Wild Mignonette

Grassland, roadsides and waste ground in the limestone areas. Quite frequent on the Bath Oolite (10). Less common elsewhere but recorded in all districts except 1 and 6.

Reseda alba L. White Mignonette

R. suffruticulosa L.

A very rare casual which was well established and abundant about Weston-super-Mare (9) from at least 1830 but which had become very scarce by 1896 (RPM) and was last reported there by Miss Livett in 1908. It was also recorded at Burnham (8) from 1836 till 1885. The only recent records, as a casual or garden escape, are from Ansford (8), 1964 (CAH), and Bedminster (10), 1977 (ALG).

VIOLACEAE

VIOLA L.

Viola odorata L. Sweet Violet

Common on hedgebanks and in woods throughout the county except on the higher parts of Exmoor. Purple-flowered plants are uncommon and often near houses where they may well be garden escapes. Of the two white-flowered varieties var. **imberbis** (Leight.) Henslow is the common plant on Jurassic soils inland from Yeovil (4) to Bath (10), and from Langport (5) and Wells (8) to the Wiltshire border. Var. **dumetorum** (Jord.) Rouy & Fouc. is more widely distributed near the coast and on less basic soils but is absent from most of the other variety's territory. Var. **subcarnea** (Jord.) Parl., with pink flowers, is rare and occurs in a number of scattered localities. Var. **sulfurea** (Car.) Rouy & Fouc. has not been seen since 1921 when it was recorded at Walton-by-Clevedon (9) by Miss Livett (**TTN**); it was known previously at Weston-super-Mare (9).

Viola hirta L. Hairy Violet

Common in grassland and wood borders on calcareous soils only, so absent from much of the Levels and from Exmoor, the Brendon and Quantock Hills. Not recorded in district 1. The ssp. **calcarea** (Bab.) E. F. Warb. (*V. calcarea* (Bab.) Gregory) is recorded on much of the Carboniferous Limestone of north Somerset. It has been suggested that this is only a late-flowering form of the species.

× **odorata** (*V.* × *permixta* Jord.). Not infrequent where the parents occur in the same area so most records are from the northern limestones, but the hybrid is also recorded from the Lias in districts 2 and 3 in V.C.5.

Viola riviniana Reichb. Common Dog-violet

V.silvatica sensu Murray

Very common on hedgebanks, in woods and on commons throughout the county but absent from much of the Levels.

Viola reichenbachiana Jord. ex Bor. Early Dog-violet

V. silvestris auct.

In woodland and on hedgebanks. As common on calcareous soils as *V. riviniana* but less common elsewhere, particularly in the west where it is rare.

× **riviniana** (*V. intermedia* Reichb., non Krock). Hybrids are to be expected in the area where the parent species grow together. As remarked by White (1912) there are many intermediates to be found but whether this is due to variability in the species or to some degree of introgression is not certain. Few examples of hybrids have been positively identified in Somerset.

VIOLACEAE

Viola canina L. Heath Dog-violet

On sand dunes at Minehead (2) and from Berrow to Brean (9) and at a few spots on Mendip heaths (9). Formerly on the peat moors (8) but not recorded there recently. There are a number of other old records in districts 3, 8, 9 and 10 and it is not clear whether the species has really declined or has been overlooked by recent recorders.

× **riviniana.** This was recorded from Worle Hill (4) by Mrs. E. S. Gregory (Marshall, 1914). It should be looked for where the parents grow together, e.g. on Mendip heathland.

× **reichenbachiana** (*V.* × *borussica* W. Becker). Recorded from Banwell Wood (9) by Mrs. E. S. Gregory (White, 1912).

Viola lactea Sm. Pale Dog-violet

Very rare, on heaths. Although more frequent in Devon this species was not detected in Somerset until Marshall found it near Crowcombe (2) in 1910. In 1915 he found it in two places in district 3; one of these is the only locality in which it has been seen recently.

× **riviniana.** Found by Marshall with the parents near Crowcombe (2) in 1910, and in district 3 in 1923 by W. D. Miller.

Viola palustris L. Marsh Violet

In bogs, marshes and by springheads, on acid soils. Common on Exmoor and the Brendons in districts 1 and 3, less so on the northern side in district 2. Common, too, on the Blackdown Hills and on the Greensand on the Wiltshire border.

Surprisingly scarce on the Quantocks and only at the northern end (2). It also occurs sparsely on the peat moors (8) and more frequently on the Old Red Sandstone of Mendip. Rarely in lowland marshes on alluvial gravel and clay, as near Congresbury (9), 1973 (JA), Churchill (9), 1976 (RSC & RGR), Hinton Blewitt (10), 1973 (JA) and Frome (10), 1971 (CAH).

Viola tricolor L. Wild Pansy

Described by Babington (1834) as frequent in cultivated ground about Bath (10), it is still found there as a weed in gardens and allotments, and very occasionally in cornfields. Elsewhere in the county it is no more than a very rare casual. A yellow-flowered plant found in 1901 in an old pasture near Exford (1) by A. Lyons was determined as *V. lutea* Huds. by Mrs. Gregory and E. G. Baker. Similar plants were found about 2 miles away near Winsford (1) in 1917 by W. D. Miller, and about halfway between these localities in 1921 by Mrs. Browning, but have not been reported since. Specimens in the herbarium at Taunton are considered to be variants of *V. tricolor*.

Viola arvensis Murr. Field Pansy

Common in cornfields and cultivated land. Rather sparse on the Levels and absent from most of Exmoor. Not recorded recently in district 1.

× **tricolor.** Very rare in Somerset because the parents seldom meet but it has been seen in neglected allotments at Bath (10).

V. × **wittrockiana** Gams (Garden Pansy) and also its hybrid with *V. arvensis* are occasionally found on rubbish dumps, etc.

POLYGALACEAE

POLYGALA L.

Polygala vulgaris L. Common Milkwort

Common in old grassland and on dry banks on calcareous soils. Rare on Exmoor and in the alluvial lowlands but recorded in all districts.

Polygala serpyllifolia Hose Heath Milkwort

Common on Exmoor and on other heaths and moors on acid soils, including the peat moors (8). Not recorded in district 5.

Polygala calcarea F. W. Schultz Chalk Milkwort

Plentiful on the chalk downs in district 7, now in the administrative county of Wiltshire. Discovered on the Oolite south of Bath in 1909 by F. Samson and still abundant on the hillside where it was first found. It was subsequently reported in three other spots within a two mile radius but has not been seen in any of them recently. An old record by W. B. Waterfall for Sandford Hill (9) was probably an error.

GUTTIFERAE

HYPERICUM L.

Hypericum androsaemum L. Tutsan

Hedgebanks and wood borders. Fairly frequent in the south and west but thinly distributed about the rest of the county and absent from the Levels.

GUTTIFERAE

Nowhere abundant, usually only one or two plants are found at any one spot. Recorded from all districts.

Hypericum inodorum Mill. Tall Tutsan

H. elatum Ait.
An occasional garden escape.

Hypericum hircinum L. Stinking Tutsan

Garden escape.

Hypericum calycinum L. Rose of Sharon

Often planted in shrubberies and gardens and naturalized in woodland rides and on roadside and railway banks. Very persistent where it has become established.

Hypericum perforatum L. Perforate St John's-wort

On hedgebanks, in calcareous grassland and in open woodland. Very common except on Exmoor.

Hypericum maculatum Crantz ssp. **obtusiusculum** (Tourlet) Hayek Imperforate St John's-wort

H. dubium Leers

Hedgebanks and rough ground. Very rare and only recorded recently at Old Cleeve (2), 1972 (CAH), Norton Fitzwarren (3), 1979 (R & IR), South Petherton (4), 1973 (CAH), Emborough (10), 1974 (R & IR), Camerton (10), 1979 (DG) and near Bath (10), 1966 (RGR). Formerly also recorded in district 9.

Hypericum tetrapterum Fr. Square-stalked St John's-wort

H. quadrangulum L.; *H. quadratum* Stokes

Common by ditches, on riverbanks and in other wet places.

Hypericum humifusum L. Trailing St John's-wort

A calcifuge species of stony banks, forestry tracks and other gravelly places. Common in the west on Exmoor, the Brendons and Quantocks. Fairly frequent on the Greensand of the eastern border and on the Old Red Sandstone of Mendip. Rare elsewhere. Much less frequent in the north of the county than appears from previous records. Recorded in all districts but it has not been seen in district 5 since 1933 when it was noted in two places near Somerton (WW & C. J. Bartlett).

Hypericum pulchrum L. Slender St John's-wort

Banks, roadsides and heaths, particularly on the acid soils of the west and south of the county. Fairly frequent on Mendip, where it is found on limestone heath as well as on the Old Red Sandstone, and in similar places on other Carboniferous Limestone exposures in the north. Although both Murray and White describe it as common throughout most of the county except in district 5 and around Bath, it is absent from the whole of the alluvial lowlands, the Polden Hills (5 & 8) and from the Oolite and Lias between Yeovil (4) and Bath (10).

Hypericum hirsutum L. Hairy St John's-wort

Hedgebanks and open woodland. Common on calcareous soils. Rare on the Levels and absent from Exmoor. Not recorded from district 1.

Hypericum montanum L. Pale St John's-wort

Rather rare in rocky and rough limestone grassland. Recently it has only been recorded in district 9 on the Carboniferous Limestone of Mendip between Brean Down, Cheddar and Compton Martin, around Weston-super-Mare and Sand Point, and between Clevedon and Bristol. It has not been seen anywhere else in the county since before 1914 but there are old records from districts 1, 2, 5 and 10. Dr Davis's statement in Warner (1802) that it was common in hedges about Bath is clearly an error.

Hypericum elodes L. Marsh St John's-wort

Elodes palustris Spach

Frequent in acid moorland bogs on Exmoor, at the northern end of Quantock (2) and on the Blackdown Hills (3, 4 & 6). Formerly on the peat moors (8) but not seen there since 1914 due to the destruction of the raised bog by peat cutting.

CISTACEAE

HELIANTHEMUM Mill.

Helianthemum nummularium (L.) Mill. Common Rock-rose

H. chamaecistus Mill.

Dry banks and hill slopes on basic soils. Very common on the Carboniferous and Oolitic limestones of north Somerset (8, 9 & 10) and frequent on the Lias of the Polden Hills and elsewhere in district 5. Also on the chalk of the eastern border (7). Rare in districts 2, 3, 4 and 6 and recorded recently only near Watchet (2), and in a few spots near Thurlbear (3), Crewkerne (4 & 6) and Yeovil (4). Absent from district 1.

Helianthemum apenninum (L.) Mill. White Rock-rose

H. polifolium Mill.

Rocky limestone slopes with a southern aspect. Only on Brean Down, where it is plentiful, and on Purn Hill, Bleadon (9). If found elsewhere it is likely to be the result of a transplant experiment (see *Arabis stricta*).

× **nummularium** (*H.* × *sulphureum* Willd.). First discovered on Purn Hill, Bleadon (9) by H. S. Thompson in 1888. It still grows there in the thin rocky limestone turf with the parents close by. It has been reported once or twice from Brean Down but if it occurs it is certainly rare there; the parent species do not grow so closely together, *H. nummularium* favouring more stable turfy areas to the rocky slope preferred by *H. apenninum*.

(**Helianthemum ledifolium** (L.) Mill. has appeared in British Floras on the strength of a statement by Hudson that it grew on "Brent Downs" but this was a mistake. Similarly **Tuberaria guttata** (L.) Fourr. (*H. guttatum* (L.) Mill.) has been erroneously credited to Somerset in some old books.)

TAMARICACEAE

TAMARIX L.

Tamarix gallica L. Tamarisk

T. anglica Webb

Sometimes planted near the sea.

CARYOPHYLLACEAE

SILENE L.

Silene vulgaris (Moench) Garcke
ssp. **vulgaris** Bladder Campion

S. cucubalus Wibel

Frequent on roadsides, field borders and in waste places over most of the county. Rather sparse on the Levels and in the south and west but recorded from all districts.

ssp. **maritima** (With.) A. & D. Löve Sea Campion

S. maritima With.

On the coast between Porlock (2) and Clevedon (9) and on Steep Holm, local and less common than formerly; not seen recently in district 8 near Burnham. It also occurs near the old lead workings on Mendip between Shipham, Priddy (9) and East Harptree (10).

× ssp. **vulgaris.** This hybrid is rare because the parent subspecies are usually ecologically separated but where they meet, as at Charterhouse-on-Mendip (9), intermediate forms occur which are presumed to be hybrids.

Silene conica L. Sand Catchfly

Very rare and only on Minehead Warren (2) where it was first found in 1894 by Miss May. It varies in quantity from year to year but seems to have spread to a wider area than it occupied when first noticed.

†**Silene gallica** L. Small-flowered Catchfly

S. anglica L. including *S. quinquevulnera* L.

A cornfield weed, always very rare in Somerset, which must now be considered extinct. The most recent records are on Butleigh Moor (5), 1955 (DHP), and Loxton Hill (9), 1951 (ER). It has also occurred as a casual in waste places near large towns but not for many years.

(**Silene nutans** L. Although shown as an introduction for square 31/76 in the *B.S.B.I. Atlas* (1962), Babington's Batheaston record (1834) can only have been a casual which had certainly gone by 1866 according to Jenyns.)

CARYOPHYLLACEAE

Silene noctiflora L. Night-flowering Catchfly

Melandrium noctiflorum (L.) Fr.

A weed of arable land still fairly frequent round Bath (10) but very rare elsewhere although formerly more widespread in north Somerset. The only recent records away from the Bath area are near Somerton and Charlton Mackrell (5), 1976 (R & IR), West Horrington (8), 1970 (JA), and Leighton (10), 1960 (Mrs. N. Wycherley). There are only three records from V.C.5, the most recent being near Washford (2) in 1937 (H. D. Badcock).

Silene dioica (L.) Clairv. Red Campion

Melandrium rubrum (Weigel) Garcke; *Lychnis dioica* L.; *L. diurna* Sibth.

Very common on hedgebanks and in woods, except in parts of the Levels and on the highest ground on Exmoor.

Silene alba (Mill.) E. H. L. Krause White Campion

Melandrium album (Mill.) Garcke; *Lychnis alba* Mill.

A common weed of arable land, roadsides and waste places over most of the county except for the higher moorland. Only once recorded in district 1, near Simonsbath, 1971 (HWB).

× **dioica.** Hybrids are not uncommon where the parents meet and are more frequent near Minehead (2) and in the eastern half of the county than elsewhere. Not all pink-flowered campions are hybrids; colour variants of both *S. dioica* and *S. alba* from red to white occasionally occur.

LYCHNIS L.

Lychnis flos-cuculi L. Ragged Robin

Common throughout the county in marshy meadows and ditches.

AGROSTEMMA L.

Agrostemma githago L. Corn Cockle.

Lychnis githago (L.) Scop.; *Githago segetum* Link

Never very common as a cornfield weed in Somerset it became increasingly scarce after 1940 and is now only a very rare casual. The only recent records are near Taunton (3), 1970 (PMH), East Coker (4), 1968 (LIS), and White Sheet Hill (7) (Wilts), 1977 (CAH).

DIANTHUS L.

Dianthus armeria L. Deptford Pink

Although both Murray and White class this species as a rare native it is doubtful whether it was ever more than a persistent garden escape in the county. It was formerly cultivated but can become a troublesome weed in gardens. Of a dozen records since 1847, mostly in the area between Yatton, Clevedon and Bristol, none have endured for long except that at Cadbury Hill, Yatton (9), where it was known for 50 years or more but has not been seen recently. The only recent records are at Selworthy (2), 1960 (TJW), and on a wall at Bruton (8) where it was first recorded in 1938 (FKM).

Dianthus gratianopolitanus Vill. Cheddar Pink

D. caesius Sm.; *D. glaucus* Huds.

On limestone rocks in and around the Cheddar Gorge (9) where it has been known for nearly 300 years. Introduced at Sand Point (9) where it was first reported in 1951. It was also introduced on the limestone rocks at Cannington (3) about 1910 but has not been seen there since 1957.

A hybrid, possibly with *D. plumarius* L., has been introduced on the rocks at Rookham, near Wells (8).

Dianthus deltoides L. Maiden Pink

Formerly in a rough sandstone pasture near Keynsham (10) where it was first recorded in 1835. By 1912 it had become very scarce and only one patch was left by 1924. It was last seen in 1928 (WDM). Collins' record (1837) "on Lias near Street" (8) was never confirmed and that for Cheddar (9) in the *Botanist's Guide* (1805) was doubtless an error. The species has also been recorded as an introduction in Croscombe Churchyard (8), where it has not persisted, and as a casual on rubbish tips at Ashton Gate (10), 1939 (C & NS), and Bedminster (10), 1979 (ALG).

VACCARIA Medic.

Vaccaria hispanica (Mill.) Rauschert Cowherb

V. pyramidata Medic.; *Saponaria vaccaria* L.

A casual on rubbish tips and waste ground.

CARYOPHYLLACEAE

SAPONARIA L.

Saponaria officinalis L. Soapwort

Hedgebanks, roadsides and waste places. Usually the double form and nearly always near houses. Widespread and in all districts except 7, but not common except on the sandy banks between Burnham (8) and Brean (9) where specimens were collected by T. Clark in 1824.

CERASTIUM L.

Cerastium arvense L. Field Mouse-ear

Very rare on calcareous and sandy soils. First detected in the county on Loxton Hill (9) in 1894 by Mrs. E. S. Gregory and later found in another spot on Mendip above Axbridge (9) in 1915 by H. S. Thompson. It is still to be found in both these localities but has failed to persist in others where it has been reported; in three different places on the limestone ridge between Clevedon (9) and Failand (10) between 1900 and 1945, on Brean Down (9), 1939 (Miss P. Catchpool), and on Lansdown (10), 1915 (Mrs. Price conf. JWW).

In 1972 it was seen for the first time in V.C.5, at Minehead Warren (2) (JVC) and it remains to be seen whether this will prove to be a permanent station.

Cerastium tomentosum L. Snow-in-summer

A garden escape which sometimes becomes naturalized on hedgebanks.

Cerastium fontanum Baumg. ssp. **triviale** (Link) Jalas
Common Mouse-ear

C. triviale Link: *C. holosteoides* Fr.; *C. vulgatum* auct.

A very common weed of grassland, arable land and waste places throughout the county.

Cerastium glomeratum Thuill. Sticky Mouse-ear

C. viscosum auct.

Very common in dry, rocky and sandy places, and on walls.

Cerastium diffusum Pers. Sea Mouse-ear

C. tetrandrum Curt.; *C. atrovirens* Bab.

In sandy places usually near the sea. Quite common on the coast near

Minehead (2) and between Burnham (8) and Clevedon (9). More rarely inland on limestone at the western end of Mendip and at Goblin Combe (9).

Cerastium pumilum Curt. Dwarf Mouse-ear

A rare plant of limestone rocks and wall tops, local but usually quite plentiful where it occurs. Quite frequent on Mendip (9), on the limestone near Weston-super-Mare (9) and in the Avon Gorge (10). It has not been recorded recently from the limestone near Clevedon (9) but it was formerly known there and may have been overlooked.

Cerastium semidecandrum L. Little Mouse-ear

Rather rare in sandy and gravelly places. More frequent on the coastal sands in districts 2, 8 and 9 than elsewhere. There are several records on Mendip from Crook Peak to Priddy (9) but apart from these the only recent inland records are from Elworthy Combe (2) and Solsbury (10).

MYOSOTON Moench

Myosoton aquaticum (L.) Moench Water Chickweed

Malachium aquaticum (L.) Fries

Riversides, ditches and marshy places. Common in the lowlands of most of the county except for the coastal areas where it is rare. Not recorded in district 1.

STELLARIA L.

Stellaria media (L.) Vill. Chickweed

An abundant weed of cultivated ground, roadsides and waste places everywhere. Very variable.

Stellaria pallida (Dumort.) Piré Lesser Chickweed

S. boreana Jord.; *S. apetala* auct.

Not infrequent on the coastal sands between Minehead (2) and Kewstoke (9). The only recent inland record is from Crook Peak (9).

Stellaria neglecta Weihe Greater Chickweed

Common on shady hedgebanks throughout the county. Recorded from all districts.

CARYOPHYLLACEAE

Stellaria holostea L. Greater Stitchwort

Very common on hedgebanks and wood borders almost everywhere but less frequent on the Levels.

Stellaria palustris Retz. Marsh Stitchwort

A rare plant of fen ditches and marshy fields in the lowland moors of districts 3, 4 and 5 and the peat moors of district 8. It was found on Nailsea Moor (9) in 1912 (Miss Sandford) but has not been seen there since 1948 (CIS).

Stellaria graminea L. Lesser Stitchwort

Field borders, rough grassland and roadsides over most of the county. Very common on acid soils but less so on calcareous ones.

Stellaria alsine Grimm Bog Stitchwort

S. uliginosa Murr.

Acid bogs and swamps, ditches and other wet places. Common on Exmoor, the Brendon, Quantock, Blackdown and Mendip Hills and on the Greensand of the eastern border. Very rare on the Levels, except for the peat moors (8), and uncommon on calcareous soils everywhere except where locally acid conditions have developed.

MOENCHIA Ehrh.

Moenchia erecta (L.) Gaertn., Mey. & Scherb. Upright Chickweed

M. quaternella Ehrh.

In thin turf on gravelly or sandy acid soils. Very rare and only seen recently near County Gate, Bossington and on Minehead Warren (2) and near Keynsham (10). Recorded in the past from a number of spots on Exmoor and the Brendon, Quantock and Blackdown Hills (2, 3 & 4) and may well survive in some of these places. It has probably gone from the old wall at Preston (4) [Rev J. Sowerby & RPM in Murray (1896)] and the very old records at Wraxall (9) (Dr A. Broughton, 1779) and Hinton Charterhouse (10) [Dr H. Gibbes in Babington (1834)] have never been confirmed.

SAGINA L.

Sagina apetala Ard. Annual Pearlwort

On wall tops, paths, rocky outcrops and other dry places. Rather common and widely distributed in all districts except 1. The usual plant is ssp. **erecta** (Hornem.) F. Hermann. Ssp. **apetala** (*S. ciliata* Fr.) is rare or overlooked; very few records were made during the recent survey but old records exist for all districts except 1, 7 & 8. The smaller variant (*S. filicaulis* auct.) has only once been reported in the county, at Brislington (10), 1931 (HST).

Sagina maritima G. Don f. Sea Pearlwort

Not uncommon but local in sandy places on the coast between Porlock (2) and Weston-super-Mare (9).

Sagina procumbens L. Procumbent Pearlwort

Very common on paths, in trampled waste places and on damp walls.

Sagina subulata (Sw.) C. Presl Heath Pearlwort

Rare, in gravelly or sandy places, on Exmoor and the Brendon, Quantock and Blackdown Hills. It was cited for V.C.6 in *Topographical Botany* on Coleman's authority and this was supported by a 1969 report from the Nature Conservancy that it had been seen behind the Huntspill sea-wall (8) but this has not been further confirmed. St Brody's record (1856) for sandy fields between Weston-super-Mare and Uphill (9) was not accepted by either Murray or White.

Sagina nodosa (L.) Fenzl Knotted Pearlwort

Sandy or gravelly places where water has lain. Very rare and apparently decreasing. Only seen recently in district 9, among the sand dunes at Berrow and Uphill and in a few spots on Mendip between Ubley and Priddy. Formerly frequent on the peat moors (8) but not seen there since 1951. There are other former records from a few scattered localities in districts 2, 3, 6, 7 and 10.

MINUARTIA L.

Minuartia verna (L.) Hiern Spring Sandwort

Alsine verna (L.) Wahlenb.

A very local plant almost confined in Somerset to the old lead workings on Mendip about Priddy and Charterhouse (9). It was recorded at Will's Neck (3) on Quantock in 1888 by H. S. Thompson but has not been seen since in that area.

CARYOPHYLLACEAE

Minuartia hybrida (Vill.) Schischk. Fine-leaved Sandwort

M. tenuifolia (L.) Hiern. non Nees ex Mart.; *Alsine tenuifolia* (L.) Crantz.

Very rare. All recent records have been near railways, working or disused, but some of them are of long standing. At Bruton (8) it has been known since 1910 and at Newton St Loe (10) since 1897. Other recent records are on the disused railway track at Yatton (9), 1974 (JA), and Sandford (9), 1975 (JA), and in the old railway quarries at Emborough (10), 1974 (R & IR). In the past it has been recorded from more natural habitats, on the Mendip limestone at Purn Hill (9), 1929 (WDM), and Crook Peak (9), 1946 (JEL), and from the Bath Oolite, most recently in 1886 (A. E. Burr). The only record from V.C.5 was near the railway at Charlton (3), 1931 (WDM).

HONKENYA Ehrh.

Honkenya peploides (L.) Ehrh. Sea Sandwort

On sandy and shingly shores, much less plentiful than formerly though it is still to be found in small quantity near Minehead (2) and at intervals between Stolford (2) and Kewstoke (9).

MOEHRINGIA L.

Moehringia trinervia (L.) Clairv. Three-nerved Sandwort

Arenaria trinervia L.

Very common on hedgebanks and in woods throughout the county except on the Levels where suitable habitats are rare.

ARENARIA L.

Arenaria serpyllifolia L. Thyme-leaved Sandwort

A common weed in arable land, gardens and waste places. Also on wall tops and in other dry places. A variety, found on coastal sand dunes, with dense inflorescences and short pedicels is var. **macrocarpa** Lloyd.

Arenaria leptoclados (Reichb.) Guss. Slender Sandwort

In similar habitats to *A. serpyllifolia* but more common, particularly on calcareous soils.

SPERGULA L.

Spergula arvensis L. Corn Spurrey

A weed of arable land on acid and sandy soils. Occurs in all districts. Common in the west and south but rather rare on calcareous soils and on the Levels.

SPERGULARIA (Pers.) J. & C. Presl

Spergularia rubra (L.) J. & C. Presl Sand Spurrey

Lepigonum rubrum (L.) Wahl.

In sandy or gravelly places on acid soils. Quite frequent on forestry tracks in the northern part of Exmoor and the Brendons (2) and occasional on the Quantocks and Blackdowns (2, 3, 4 & 6). Very rare in North Somerset, V.C.6, the only records being near Keynsham (10), 1915 (C. Bucknall) and near Yarnfield (7 & 10) (Wilts), 1957 (D. Grose). It has been seen recently in both of these localities.

Spergularia rupicola Lebel ex Le Jolis Rock Sea Spurrey

S. rupestris Lebel

Very rare and only on the cliffs near Hurlstone Point (2) where it was first found by Marshall in 1907. It was reported on a lawn at Anchor Head, Weston-super-Mare (9) in 1963 (RMB) but this must have been a casual occurrence as it has not been seen there since and has never been found on the rocks in that area. A specimen in **TTN** dated 1869 collected at Clevedon (9) by H. F. Parsons has proved to be *S. marina.*

Spergularia media (L.) C. Presl Greater Sea Spurrey

S. marginata Kittel; *Lepigonum marginatum* Koch

Common in muddy salt marshes along the coast and up the tidal estuaries of the Parrett, Brue, Axe and Avon.

Spergularia marina (L.) Griseb. Lesser Sea Spurrey

S. salina J. & C. Presl; *Lepigonum salinum* Fr.

In drier parts of salt marshes than *S. media* and just as common. Occasionally higher up the shore but not far from salt water.

CARYOPHYLLACEAE

Corrigiola L.

(**Corrigiola litoralis** L. Although shown in the *B.S.B.I. Atlas* (1962) as an introduction in square 31/76 the species has only once been recorded there, as a casual, in 1896 by S. T. Dunn).

Herniaria L.

†**Herniaria glabra** L. Smooth Rupturewort

Sole's record of this (1791) on the coast at Weston-super-Mare (9) has been regarded as an error but it may well have occurred there once. It was seen there as a casual in 1946 (G. Nicholls).

Scleranthus L.

Scleranthus annuus L. Annual Knawel

A weed of arable fields and poor pastures on acid and sandy soils. Formerly quite widespread and recorded in all districts except 1 and 5 but now very rare. The only recent records are from Yenworthy (2), 1962 (MT), at Hinton Charterhouse (10), 1978 (AD), and in a long-established locality on the sandstone near Keynsham (10), 1978 (ALG).

†**Scleranthus perennis** L. Perennial Knawel

The only evidence that this very rare plant once grew in Somerset rests on a very poor specimen at Taunton, collected by Gapper about 1825 at Spaxton (3), and on another from Warleigh Common (10), formerly in the Bath herbarium but which cannot now be found, which was considered by Murray and E. F. Linton doubtful but possibly correctly named.

PORTULACACEAE

Montia L.

Montia fontana L. Blinks

Wet places by streams and springs and in marshy fields on acid soils. Common on Exmoor and the Brendon, Quantock and Blackdown Hills. Rather rare in North Somerset except on the Greensand of the eastern border (7 & 10), the peat moors near Shapwick (8) and at Weston-in-Gordano, the Old Red Sandstone of Mendip (8, 9, & 10) and the Pennant Sandstone near Keynsham and Clutton (10). Not recorded from district 5.

The distribution of the subspecies has not been fully worked out. The common plant of the western moorland and the Mendip bogs is ssp. **amporitana** Sennen. Ssp. **chondrosperma** (Fenzl) Walters prefers drier and less acid places and has been found on the peat moors (8) and near Keynsham (10). Ssp. **variablis** Walters has only been identified once in the county, on the peat moors (8), 1912 (C. E. Moss).

Montia perfoliata (Willd.) Howell — Spring Beauty

Claytonia perfoliata Donn ex Willd.

First recorded in the county as a garden weed at Clevedon (9) in 1912 (Miss M. A. G. Livett) and five or six times since then as a casual in gardens or waste places, it has become well established on the sand dunes near Dunster (2) since 1960 (Miss P. P. Hill) and at Brean (9) since 1968 (R & IR).

Montia sibirica (L.) Howell — Pink Purslane

Claytonia alsinoides Sims

A garden escape which can become well established and spread by streams and in damp woods. First recorded in Somerset as a weed in Whitelackington Churchyard (4) (Preb J. Hamlet) in 1926 and as a casual at Somerton (5) and Street (8), 1943/44 (E. F. Payne), it now appears to be well established by the Exe between Bridgetown and Barlynch (1), around Luccombe and Tivington (2), at Wrangway near Wellington (3), near Churchstanton (6) (V.C.3) and near Kilmington (7) (Wilts).

PORTULACA L.

Portulaca oleracea L.

A rare casual, first recorded in the county at Bridgwater (3), 1954 (EJH).

AMARANTACEAE

AMARANTHUS L.

Amaranthus retroflexus L. — Common Amaranth

The most frequent species of this genus of waste ground and rubbish-tip casuals. Recent records are at Bridgwater Docks (3 & 5), 1967-75 (HWB), Kilminton (7) (Wilts), 1977 (PT), Batcombe (10), 1964 (CAH), Gare Hill (10), 1973 (PT), and Brislington (10), 1978 (ALG).

AMARANTACEAE

Amaranthus hybridus L. Green Amaranth

The type of this species is a rare casual recorded in the Bristol area (10) in 1922 (C & NS) and 1979 (ALG), and at Bridgwater Docks (3) in 1959 (EJH). The ssp. **incurvatus** (Timeroy ex Gren. & Godr.) Brenan (*A. cruentus* L.) was first recorded in the county on a tip at Ashton Gate (10), 1935 (CIS & JPMB), and was found on a tip at Brislington (10) in 1978 (ALG).

Amaranthus albus L.

An infrequent casual. Recent records are from waste ground at Bath (10), 1963 (RGR), and a rubbish tip at Brislington (10), 1978 (ALG).

Amaranthus blitoides S. Watson

A species from North America, first recorded in the county at Ashton Gate (10), 1940 (CIS & JPMB det. Kew) and seen on the Brislington tip (10), 1979 (ALG det. EJC).

Amaranthus standleyanus Parodi ex Covas

A. vulgatissimus auct.

A rare casual in Britain, recorded at Ashton Gate (10), 1932 (CIS), and more recently at Brislington (10), 1978 (ALG, CML & AT det. EJC).

CHENOPODIACEAE

CHENOPODIUM L.

Chenopodium bonus-henricus L. Good King Henry

Waysides and waste places near farms and in villages where it was formerly cultivated. Long persistent in many of its localities. Not common but widely distributed. Recorded in all districts but not seen recently in 7.

Chenopodium polyspermum L. Many-seeded Goosefoot

A frequent weed of arable land and waste places which seems to be increasing in the lowland areas. Recorded in all districts but rare on Exmoor and other higher ground and around Bath (10).

†**Chenopodium vulvaria L.** Stinking Goosefoot

Persisted at Bath (10) under walls by the old Gas Works for nearly a century, disappearing about 1930 (WDM). Not seen elsewhere since 1914 [Portishead (9) (CIS)] but there are also old records of its occurring as a casual at Minehead (2), Shapwick (8), Burnham (8) and Berrow (9).

Chenopodium album L. Fat Hen

An abundant and very variable weed of cultivated ground and waste places. The closely-allied species **C. suecicum** J. Murr. (*C. album* var. *viride* auct.) is believed to occur also. The plant with red-striped stems from Twerton (10) referred to by White (1912) has been identified from his specimen in the Liverpool herbarium as ssp. **striatum** (Krasan) J. Murr. by J. P. M. Brenan.

Chenopodium ficifolium Sm. Fig-leaved Goosefoot

Although classed by both Murray and White as rare this species is now a very common weed of cultivated lands and manure heaps throughout the Levels and other lowland areas. Still rare elsewhere, particularly in the west, but recorded from all districts.

†**Chenopodium murale** L. Nettle-leaved Goosefoot

Formerly recorded from waste places and cultivated land in several places along the coast from Minehead (2) to Portishead (9) and from a few further inland. Last seen in the county in two spots on Loxton Hill (9) in 1951 (ER).

†**Chenopodium urbicum** L. Upright Goosefoot

Although Babington said (1834) that it was frequent about Bath (10) and it was seen there as late as 1893 this has always been very rare in Somerset and has only been recorded twice since 1907, in both instances (1923 and 1940) near Taunton (3) (WW).

†**Chenopodium hybridum** L. Maple-leaved Goosefoot

Formerly in waste places about Bath (10) but not seen there since 1896 (D. Fry). Old records for Dunster (Coleman, 1849) and Weston-super-Mare (9) (St Brody, 1856) have never been confirmed since.

Chenopodium rubrum L. Red Goosefoot

On manure heaps and rich waste ground. Common throughout the lowlands. Rather rare on higher ground and in the west but recorded in all districts.

CHENOPODIACEAE

†**Chenopodium glaucum** L. Oak-leaved Goosefoot

A very rare casual that has seldom been recorded in the county. The only records are Brean Down Farm (9), 1904-06 (CEM), Highbridge (8), 1938 (FKM), near Brent Knoll station (9), 1937 (Mrs. Bell & HJG), 1938 (ER & Miss A. G. Miller), and the Bristol area (10), 1917-30 (CIS).

Chenopodium ambrosioides L.

A rare casual the only Somerset records for which are at Twerton (10), 1896 (S. T. Dunn), Yeovil (4), ca. 1959 (VIR) and Brislington (10), 1979 (TGE & ALG conf. EJC).

Chenopodium probstii Aellen

A wool alien recorded in the county on rubbish tips at Yeovil (4), ca. 1959 (VIR) and Windsor Hill, Shepton Mallet (8), 1970 (CAH det. JEL).

BETA L.

Beta vulgaris L. ssp. **maritima** (L.) Arcang. Sea Beet

Common along the whole length of the coast from Porlock (2) to Portbury (10) on muddy and shingly shores and up the tidal estuaries.

ATRIPLEX L.

Atriplex littoralis L. Grass-leaved Orache

Very rare and uncertain in its appearance on the shore between Burnham (8) and Kewstoke (9). St Brody's record for Weston-super-Mare (9) was not accepted by Murray but the presence of the plant in Somerset was definitely established in 1889 by H. S. Thompson when he found it near the mouth of the Brue below Burnham. It was not seen again till 1946 when it was recorded south of Brean (9) (Cdr. R. D. Graham) and it has been reported there more recently by several observers. Other recent records are Stert I. (2), 1975 (RSC), Berrow (9), 1976 (RSC), Uphill (9), 1963 (RMB), and Sand Bay (9), 1970 (ESS).

Atriplex patula L. Common Orache

A. erecta Huds.

Very common in cultivated land and waste places throughout the county except on the higher parts of Exmoor.

Atriplex prostrata Boucher ex D.C. Spear-leaved Orache

A. deltoidea Bab.; *A. microsperma* sensu Bab.; *A. triangularis* Willd.; *A. hastata* auct.

A very common and very variable weed of cultivated ground, waste places, salt marshes and sea shores. Rarer on higher ground and not recorded in district 1.

Atriplex longipes Drejer

The presence of this coastal species in the British Isles was first confirmed in 1975 by a find in south-west Scotland [P. M. Taschereau (**MANCH**)] but specimens from Burnham (8) [B. A. Hulme (**LIV**)] and Brean (9) [E. M. Jones (**OXF**)] examined by Taschereau were said by him to be strongly reminiscent of *A. longipes* but not sufficient to permit of certain identification. It has since been found on the Gloucestershire side of the River Avon (1977, IFG conf. PMT).

Atriplex glabriuscula Edmondst. Babington's Orache

A. babingtonii Woods

Common on sand and shingle on the coast from Porlock (2) to Portishead (9) but less so than *A. prostrata.*

Atriplex laciniata L. Frosted Orache

A. sabulosa Rouy

Very rare and uncertain in its appearance on sandy shores. Old records by Collins from Steart (2) and Burnham (8) and by St Brody from Weston-super-Mare (9) were not accepted by Murray but these botanists were vindicated by finds at Steart and Burnham by Marshall in 1907 and 1906 respectively, and at Sand Bay (9) by Mrs. Sandwith in 1915. The only recent records are from Stert Island (2), 1975, Berrow (9), 1976, and Uphill (9), 1976, all by RSC.

Atriplex hortensis L. Garden Orache

An uncommon garden escape.

HALIMIONE Aellen

Halimione portulacoides (L.) Aellen Sea Purslane

Atriplex portulacoides L.; *Obione portulacoides* (L.) Moq.

CHENOPODIACEAE

A rare plant of the muddy shores round the mouth of the Parrett in districts 2, 3 and 8. Elsewhere on the coast the only recent record was at Porlock Dock (2), 1972 (CAH), though formerly it was recorded at Minehead (2) (Collins, 1837) and near Clevedon (9), where it was last seen in 1905.

SUAEDA Forsk. ex Scop.

Suaeda maritima (L.) Dumort. Annual Sea-blite

Lerchia maritima O. Kuntze

Common in salt marshes and on muddy shores of the coast and river estuaries from Porlock (2) to the mouth of the Avon (10).

†**Suaeda vera** J. F. Gmel. Shrubby Sea-blite

S. fruticosa auct.; *Lerchia obtusifolia* Steud.

M. de l'Obel (Lobelius) said he picked this plant on his visit to Steep Holm about 1569 but both Murray and White, who visited the island together in 1891, regarded this as almost certainly an error. They recorded no strand plants on the shingle beach and none grow there now, but several such species were noted by earlier visitors. There have probably been extensive natural changes to the configuration of the beach over the years, apart from attempts to make an artificial harbour, and it is not unlikely that de l'Obel's find was correctly identified.

SALSOLA L.

Salsola kali L. Prickly Saltwort

A rare plant of sandy shores between Minehead (2) and Kewstoke (9). Formerly on Steep Holm, 1883 (J. Storrie), and at Portishead (9), 1906 (JWW).

SALICORNIA L. (Glasswort)

The practical difficulties of identifying the various species of **Salicornia** have made it impossible to define their distribution in Somerset. They occur on the mud flats and in the lower zones of salt marshes all along the coast from Porlock (2) to Portbury (10). Usually referred to under the aggregate names of **S. herbacea** L. or **S. stricta** Dumort., the species present include **S. ramosissima** Woods, **S. europaea** L., and **S. dolichostachya** Moss.

(**Arthrocnemum perenne** (Mill.) Moss (*Salicornia perennis* Mill.) was incorrectly credited to Somerset in the *Botanist's Guide* (1805) where Sole is quoted as the authority. Both Murray and White considered this a mistake, but reference to Sole's list in Collinson (1791) shows that the plant he recorded "In the salt marshes near Highbridge" was *S. europaea*.)

KOCHIA Roth

Kochia scoparia (L.) Schrad. Summer Cypress

Garden escape.

TILIACEAE

TILIA L.

Tilia platyphyllos Scop. Large-leaved Lime

Not known in Somerset except as a planted tree until 1973 when P. J. M. Nethercott found a group of five in an undisturbed rocky area in Leigh Woods on the slopes of the Avon Gorge (10). The trees are probably more than 100 years old and the site is similar to those in the Wye valley where the species is undoubtedly native.

Tilia cordata Mill. Small-leaved Lime

Quite frequent in woods on the Carboniferous Limestone of Mendip, Broadfield Down and the ridge between Weston-in-Gordano and Bristol in districts 9 and 10. Occasional in woodland on the Lias, as at Orchard Portman and Fivehead (3) and Compton Dundon (5). Elsewhere only as a planted tree.

× **platyphyllos** Common Lime

T. × *vulgaris* Hayne: *T.* × *europaea* auct.

Commonly planted in all parts of the county.

MALVACEAE

MALVA L.

Malva moschata L. Musk Mallow

In rough dry grassland, on hedgebanks and by roadsides. Common over much of the county but scarce on the Levels and absent from the higher ground of Exmoor. Recorded in all districts.

MALVACEAE

Malva sylvestris L. Common Mallow

including *M. ambigua* Guss.

Common on roadsides and in waste places except on Exmoor and higher ground elsewhere. Not recorded recently in district 1.

Malva neglecta Wallr. Dwarf Mallow

M. rotundifolia auct.

By farmyards, on roadsides and in dry waste places. Not common but recorded in all districts except 1 and not seen recently in 6 or 7.

Malva parviflora L. Least Mallow

A rare casual, only recorded recently at Brislington (10), 1978 (ALG).

LAVATERA L.

Lavatera arborea L. Tree Mallow

On maritime rocks and in sandy waste places near the sea between Porlock Weir (2) and Clevedon (9). Occasionally as a garden escape further inland. In most of its localities it probably originated as an escape from cultivation but it may well be native on the cliffs of Steep Holm and Brean Down (9). It has been on Steep Holm since before 1830 and is still plentiful there.

Lavatera trimestris L.

An uncommon garden escape.

ALTHAEA L.

Althaea officinalis L. Marsh Mallow

A rare plant of marsh ditches near the sea only seen recently at Steart (3), in the lowlands between Weston Zoyland (5) and Burnham (8), and on the coast near Redcliffe Bay (9).

Althaea hirsuta L. Hairy Mallow

Found in 1875 in an open stony pasture in Copley Wood (5) by J. G. Baker who considered it to be a true native there. This opinion was accepted by Murray and some other botanists but its true status is doubtful. It proved very persistent but has not been seen in the original area since 1954. In 1950 it was found in a similar habitat near Aller (5) some six miles away by A. D. & O. M. Hallam and it still survives there. It was exceptionally abundant in 1977 after the severe drought of 1976 which reduced competition from grasses and other species but reverted to a more usual population of only a few plants the following year. Elsewhere it has been found occasionally as a casual.

ALCEA L.

Alcea rosea L. Hollyhock

Althaea rosea (L.) Cav.

Garden escape.

LINACEAE

LINUM L.

Linum bienne Mill. Pale Flax

L. angustifolium Huds.

Dry calcareous grassland. Quite frequent on the Lias near the coast between Minehead and Steart (2), on the Polden Hills (5 & 8), on Mendip between Wells (8) and Uphill (9), on Worle Hill (9) and between Clevedon and Portishead (9). A casual elsewhere. Not recorded in districts 1, 6 and 7.

Linum usitatissimum L. Flax

A not infrequent casual.

(**Linum perenne** L. (*L. anglicum* Mill.). The record by Collins (1837) from Puriton (8) was probably an error for *L. bienne* which occurs there.)

Linum catharticum L. Fairy Flax

Dry grassland. Very common on calcareous soils. Less so elsewhere but recorded from all districts.

RADIOLA Hill

Radiola linoides Roth All-seed

Very rare on bare damp peaty or sandy ground. Only seen recently on the Shapwick peat moor (8) but recorded formerly near Holford (2) and in a few spots on the Blackdown Hills (3, 4 & 6).

GERANIACEAE

Geranium L.

Geranium pratense L. Meadow Cranesbill

Common by roadsides and in meadow borders in the eastern half of the county. Rare elsewhere and though recorded in all districts it is only a garden escape in most of its stations in the west.

Geranium endressii Gay French Cranesbill

Garden escape.

Geranium versicolor L. Pencilled Cranesbill

An occasional garden escape thoroughly well naturalized near Selworthy (2) and Goathurst (3). It was also known near Bourton Combe (9) for over fifty years but has not been seen there since 1936.

Geranium phaeum L. Dusky Cranesbill

Another garden escape which has become well naturalized in a number of places including Cutcombe (2), East Coker (4), Cole (8), Charterhouse-on-Mendip (9) and Gare Hill (10) (Wilts). In other places where it was formerly recorded it has not persisted.

Geranium sanguineum L. Bloody Cranesbill

A very rare plant of limestone rocks now only known at Brean Down, Cheddar Gorge and Ebbor Rocks, all in district 9. It formerly grew on the Somerset side of the Avon Gorge (10) but has not been seen there since 1882. Other old records are from Brockley Combe (9) (W. B. Waterfall) and from a locality near Clevedon (9) which was destroyed by building. Also seen as a garden escape occasionally and another subject of the Birstol University transplant experiment (see *Arabis stricta*).

Geranium pyrenaicum Burm.f. Hedgerow Cranesbill

Hedgebanks, roadsides and field borders. Described by Murray as rather rare it appears to be increasing its range and is now recorded in all districts except 7. Very common round Bath (10) and now quite frequent in many other districts.

Geranium columbinum L. Long-stalked Cranesbill

Dry grassland and banks. Common on calcareous soil but not confined to it. Recorded in all districts.

Geranium dissectum L. Cut-leaved Cranesbill

A very common weed of cultivated and waste land throughout the county.

Geranium rotundifolium L. Round-leaved Cranesbill

On roadsides, under hedgebanks and walls and in waste places. Quite common in the north of the county in districts 9 and 10, particularly round Bath and Bristol. A rare casual elsewhere. The only records from V.C.5 are near Wellington (3), 1973 (HWB), East Coker (4), 1972 (CJC), and Barwick (4), 1974 (CAH).

Geranium molle L. Dovesfoot Cranesbill

A very common weed of cultivated land, waste places and poor pastures.

Geranium pusillum L. Small-flowered Cranesbill

Cultivated land, waste places and roadsides. Rather rare but widely distributed and recorded from all districts except 1, 6 and 7.

Geranium lucidum L. Shining Cranesbill

On walls, hedgebanks and limestone rock screes. Common in the north and east and in the vales of the west and south. Rare on the Levels and absent from the higher parts of Exmoor.

Geranium robertianum L. ssp. **robertianum** Herb Robert

Very common on hedgebanks, in woods, on walls and in waste ground.

ssp. maritimum (Bab.) H. G. Bak.

A maritime plant growing on shingle, cliffs and walls near the sea. Recorded recently only at Porlock and Minehead (2) and on the cliffs of

Steep Holm. Formerly recorded also at Steart (2) and Worle (9) and it may well still exist in these areas.

Geranium purpureum Vill. Little Robin

A very rare plant of rocky places, usually near the sea, recorded in Somerset only from the Cheddar (9) and Avon (10) Gorges. Varieties of *G. robertianum* were formerly confused with this species and some of the old records were misidentified.

ERODIUM L'Hérit.

Erodium maritimum (L.) L'Hérit. Sea Storksbill

Rare and local on sandy or rocky ground near the sea and on limestone rocks further inland. Recorded recently on the coast only near Bossington (2), on Brean Down, Worle Hill and Sand Point (9), and on Steep Holm where it is abundant. Formerly also at Minehead and Dunster (2), at Berrow and near Clevedon (9). Inland only on Mendip in a few spots from Cheddar Gorge to Wavering Down (9). Not seen recently on the Somerset side of the Avon Gorge (10) where it was first recorded by Merrett in 1666, nor on the limestone of Broadfield Down where White said it was abundant in some years.

Erodium moschatum (L.) L'Hérit. Musk Storksbill

Rare in sandy waste places near the sea and occasionally as a casual inland. Whether this species is native in Britain is still in doubt but in Somerset it has certainly been long established at Minehead (2) and Purn Hill, Bleadon (9). It also occurs at Steart (3) and Berrow (9). It has not been recorded elsewhere recently except near Sampford Brett (2), 1971 (HWB), and Brislington (10), 1979 (ALG). Formerly it was also recorded as a possible native at some spots on Mendip between Cheddar and Bleadon, at Weston-super-Mare and on the ridge east of Clevedon (9) as well as in a number of places in districts 4, 5 and 9 as an obvious casual. The small early-flowering form at Purn Hill and other places in shallow soil on Carboniferous Limestone was considered by Marshall to be a distinct variety (var. **minor** Rouy) but in 1928 W. D. Miller said that after cultivation for two years in a garden it had reverted wholly to type.

Erodium cicutarium (L.) L'Hérit. Common Storksbill

In dry, especially sandy, waste places, in thin rocky turf and on wall tops.

Frequent along the coast and on Mendip but rather rare elsewhere. Not recorded in districts 1, 6 or 7. The very glandular maritime ssp. **bipinnatum** Tourlet (*E. glutinosum* Dumort.) may occur on the coastal sand dunes but has not been positively identified. Plants intermediate between this and ssp. **cicutarium** have been called ssp. **dunense** Andreas and such plants have been found at Minehead (2) and Kewstoke (9).

The wool adventives **Erodium botrys** (Cav.) Bertol and **Erodium obtusiplicatum** (Maire, Weiller & Wilczek) J. T. Howell, were recorded on a tip at Glastonbury (8) in 1970 (CAH & JGK det. JEL).

TROPAEOLACEAE

Tropaeoleum L.

Tropaeoleum majus L.

Garden escape.

OXALIDACEAE

Oxalis L.

Oxalis acetosella L. Wood Sorrel

Woods and shady hedgebanks. Very common over much of the county and recorded in all districts but surprisingly rare on the Levels and in the lowlands of districts 3, 4 and 5.

Oxalis corniculata L. Yellow Sorrel

A frequent garden escape which becomes established on paths and pavements and can be a troublesome weed.

Oxalis europaea Jord. Upright Yellow Sorrel

O. stricta auct.

A weed in gardens and allotments which quite often escapes.

Oxalis articulata Savigny Pink Oxalis

O. floribunda Lehm.

Often cultivated in gardens and a common escape which becomes more or less naturalized in waste ground and by roadsides.

OXALIDACEAE

Oxalis incarnata L.

A garden escape established in waste places near Minehead (2) and elsewhere in the west.

BALSAMINACEAE

IMPATIENS L.

(**Impatiens noli-tangere** L. Old records from Prior Park, Bath, and Vallis (10) and a more recent one from Cannington (3), 1922 (WDM), were probably garden escapes.)

Impatiens capensis Meerb. Orange Balsam

I. biflora Walt.

An alien from North America, first found in the county near Frome in 1902 by Miss S. C. Harding, which has become established in a few places by the Rivers Frome and Avon and some tributary streams (10).

Impatiens parviflora DC. Small Balsam

Another alien which has become firmly established since 1943 in Bourton and Brockley Combes (9) and has been recorded as a casual in a few other places in districts 9 and 10. Some early records for this species proved to be errors for *I. capensis.*

Impatiens glandulifera Royle Indian Balsam

Originally a garden escape it is now naturalized on the banks of rivers and streams throughout the county in all districts. It was first recorded in the county about 1920 by a stream at Cannington (3) by E. J. Hamlin, but its main spread appears to have started about 1939.

ANACARDIACEAE

RHUS L.

Rhus typhina L. Stagshorn Sumach

A garden escape established on walls and in waste places especially in the limestone districts.

ACERACEAE

ACER L.

Acer pseudoplatanus L. Sycamore

Very common everywhere in woods, hedges, gardens, etc. Although originally introduced it is now completely naturalized and a positive weed in woodland and anywhere it is allowed to seed itself. On Steep Holm, where it was first recorded in 1883, its aggressive spread during the last thirty years on the slopes favoured by the Peony has caused the latter's sad decline and threatens its survival.

Acer platanoides L. Norway Maple

Increasingly planted of recent years and sometimes found self-sown away from plantations.

Acer campestre L. Field Maple

Very common in hedges and woods, particularly on calcareous soils, almost everywhere except in the extreme west on Exmoor where it is rare.

HIPPOCASTANACEAE

AESCULUS L.

Aesculus hippocastanum L. Horse Chestnut

Commonly planted but nowhere naturalized.

AQUIFOLIACEAE

ILEX L.

Ilex aquifolium L. Holly

In woods and hedges. Often planted so recorded in a high proportion of tetrads except those on the Levels. More common as a native on acid soils than on calcareous ones.

CELASTRACEAE

EUONYMUS L.

Euonymus europaeus L. Spindle Tree

Common in hedges, downland scrub and woods, particularly on calcareous soils, over most of the county except on Exmoor and the Levels.

BUXACEAE

BUXUS L.

Buxus sempervirens L. Box

Often planted in woodland and in hedges. It has been suggested that some old bushes on limestone cliffs in Bourton Combe (9) were native but they were probably birdsown and arose from plantations not far away.

RHAMNACEAE

RHAMNUS L.

Rhamnus catharticus L. Buckthorn

In hedges and woods particularly on a calcareous soil. Not uncommon in the eastern half of the county. Absent from the western half except on the Lias near the coast east of Minehead (2). Not recorded in district 1 nor recently in 3.

FRANGULA Mill.

Frangula alnus Mill. Alder Buckthorn

Rhamnus frangula L.

Hedges and woods on damp peaty soils. Frequent on the peat moors (8) and on the Blackdown Hills (3, 4 & 6). Rare elsewhere. Not recorded in district 2 and not recently in 1 or 7.

VITACEAE

VITIS L.

Vitis vinifera L. Grape Vine

An occasional casual.

PARTHENOCISSUS Planch.

Parthenocissus quinquefolia (L.) Planch. Virginia Creeper

Escape from cultivation occasionally naturalized in waste places and on roadsides.

LEGUMINOSAE

LUPINUS L.

Lupinus arboreus Sims Tree Lupin

An uncommon garden escape.

LABURNUM Fabr.

Laburnum anagyroides Medic. Laburnum

Frequently planted and sometimes spreads to railway banks, roadsides and waste places.

GENISTA L.

Genista tinctoria L. Dyer's Greenweed

Frequent in poorly drained pastures on Jurassic and Triassic clays of the eastern half of the county and on the Blackdown Hills (3, 4 & 6). Very rare elsewhere. The only recent record in district 2 is near Lilstock, 1971 (HWB), and it has never been recorded in district 1.

Genista anglica L. Petty Whin

Very rare on rough damp moors and heaths. Only seen recently near East Anstey (1), on moors between Bathealton and Milverton (3) and at Buckland Wood and Blagdon Hill on the northern slope of the Blackdown Hills (3). It may occur elsewhere on the Blackdowns in districts 4 and 6. Surprisingly rare on Exmoor; the only record was near Langcombe Head (2), 1938 (NGH). Formerly recorded near Chard (4 & 6) and below Blackslough Woods (8) but not seen in either locality for very many years. An 1883 record from Steep Holm was almost certainly an error.

ULEX L.

Ulex europaeus L. Gorse

Very common on heaths, moors and downs over much of the county.

Rather rare on the Levels, except on the peat moors (8), and in the north-east near Bath and Frome (10).

× **gallii.** This hybrid is probably frequent where the parent species grow together as they are interfertile but the flowering periods do not overlap much. It is difficult to detect due to its variability but it is intermediate in characters and flowering periods between the parents. It has tentatively been identified on Crook Peak (9) and near Pensford (10).

Ulex gallii Planch. Western Gorse

Very common on heaths and moorlands on the acid soils of Exmoor and the Brendon, Quantock and Blackdown Hills. It is also quite frequent on the Carboniferous Limestone of Mendip and Broadfield Down where leaching has produced locally acid conditions, as well as on the Old Red Sandstone of Mendip, on the Coal Measures near Pensford (10) and on the eastern Greensand (7, 8 & 10).

(**Ulex minor** Roth (*U. nanus* T. F. Forst.) was recorded by Murray along the southern border of the county but it is now considered that the species does not occur as far west as Somerset and that the plants recorded must have been small examples of *U. gallii.*)

CYTISUS L.

Cytisus scoparius (L.) Link Broom

Sarothamnus scoparius (L.) Wimm. ex Koch; *S. vulgaris* sensu White

On dry sandy, gravelly and rocky soils. Calcifuge so common in the west (1, 2 & 3), on the Greensand in the east (7 & 10) and on the sandstones and Coal Measures in the north (9 & 10). Rare elsewhere but sometimes planted on roadside banks. Not recorded, except as an introduction, in district 5.

ONONIS L.

Ononis repens L. Common Restharrow

Fairly common in rough calcareous grassland, on roadsides and in sandy places near the sea where a spinous form, the var. **horrida** Lange, is frequent. Not recorded in districts 1 or 6.

Ononis spinosa L. Spiny Restharrow

In rough pastures, preferring a clay soil. Fairly common in the eastern half of the county. Very rare in the west except on the Lias near the coast of district 2; the only recent records west of the Quantock Hills are near Stogumber (2), 1973 (CAH), and near Milverton (3), 1971 (HWB).

MEDICAGO L.

Medicago sativa L.

ssp. **sativa** Lucerne

A widespread escape from cultivation on roadsides and in waste places. Recorded from all districts except 1.

ssp. **falcata** (L.) Arcangeli (*M. falcata* L.) Sickle Medick

A rare casual only seen recently in district 10, at Englishcombe, 1963 (RGR), Brislington, 1971 (AFD), and Bedminster, 1979 (ALG).

× ssp. **sativa** (*M* × *varia* Martyn) has been recorded in V.C.6 a few times in the past but the only recent record was on a tip at Bedminster (10) in 1979 (ALG).

Medicago lupulina L. Black Medick

Abundant in grassland, on roadsides and in waste places in all parts of the county except on Exmoor where it is rather rare.

Medicago minima (L.) Bartal. Bur Medick

A very rare casual, the only recent record being at Wookey (9), 1977 (CAH). In *Flora Bathoniensis* Babington quotes a record for Weston Fields near Bath (10) by Dr Davis which Murray dismisses as a mistake. The error was Babington's as Davis's original record (in Warner, 1802) was for *M. polymorpha.*

Medicago polymorpha L. Toothed Medick

M. hispida Gaertn.; *M. denticulata* Willd.

Very rare and possibly native on sandy ground near the sea at Minehead Warren (2) where it was found in 1967 by J. J. Robins (**TTN**). Formerly also near Dunball (5) but not seen there for many years. A rare casual elsewhere only recorded recently on mill tips at Wellington (3), 1978, and Glastonbury (8), 1970 (both CAH & CES), but formerly more widespread.

LEGUMINOSAE

Medicago arabica (L.) Huds. Spotted Medick

By roadsides, in waste places and sometimes in arable land. Quite common near the sea from Porlock (2) to Portbury (10) and inland in the lowlands of districts 3, 4, 5 and 8 and up the Avon valley as far as Keynsham (10). A rare casual elsewhere. Not recorded in districts 1, 6 or 7.

Medicago arborea L. Tree Medick

A Mediterranean shrub naturalized on the cliff at Clevedon (9), 1973 (IFG).

MELILOTUS Mill.

Melilotus altissima Thuill. Tall Melilot

M. officinalis sensu Murray & White

Roadsides, waste places and field borders. Common over much of the county except on high ground. Rather rare in the west except near the coast. Not recorded in districts 1 or 6.

Melilotus officinalis (L.) Pall. Ribbed Melilot

M. arvensis Wallr.

A casual of waste ground particularly near railways and docks. Fairly frequent near Bristol (10). Records elsewhere are scattered in all districts except 1 and 6.

Melilotus alba Medic. White Melilot

Originally introduced. Now an increasing weed of waste places quite persistent in some. Not yet frequent but recorded in all districts except 1, 6 and 7.

Melilotus indica (L.) All. Small Melilot

A rare casual of waste ground recently recorded in districts 3, 4, 5, 8 and 10 and formerly in 2 and 9 also.

Melilotus sulcata Desf. French Melilot

A bird-seed alien first recorded in the county at Bedminster (10), 1930 (CIS).

TRIGONELLA L.

Trigonella caerulea (L.) Ser.

A rare casual formerly recorded in waste places in several localities near Bristol and Bath but only seen once since 1912, at Bath (10) on a rubbish tip, 1978 (DG).

TRIFOLIUM L.

Trifolium ornithopodioides L. Fenugreek

Trigonella ornithopodioides (L.) DC.; *T. purpurascens* Lam.

Rather rare and inconspicuous in thin turf in sandy or gravelly places near the coast between Porlock and Steart (2), by the Parrett at Combwich (3), at Brean Down and Sand Point (9). Formerly at Burnham (8) but not seen there since 1938.

Trifolium pratense L. Red Clover

Abundant in fields, on roadsides and in waste places throughout the county both wild and as a relic of cultivation.

(**Trifolium ochroleucon** Huds. W. Christy's record of *T. squamosum* at Banwell (9) was mistakenly entered in the *New Botanist's Guide* (1835) under this name.)

Trifolium medium L. Zigzag Clover

Wood borders, roadsides and pastures. Rather common over much of the county, less so in the lowlands and very rare on Exmoor. Not seen recently in district 1.

Trifolium squamosum L. Sea Clover

Rare and local in salt marshes and grassland near the sea and tidal river estuaries from Stolford (2) to Clevedon (9). Formerly recorded further west along the coast as far as Porlock (2) and in the north near Portishead (9) but not seen recently in either of these areas. Two inland localities in district 8, on Yarley Hill and at Barton St David, were thought by Murray to be relics of the times when Glastonbury Moor was still an arm of the sea (*J. Bot.*, 1882). The plant has gone from other inland stations near Cossington (8) and Banwell (9). An occurrence at West Monkton (3) in 1923 (WW), was probably casual.

LEGUMINOSAE

†**Trifolium stellatum** L. Starry Clover

Recorded twice as a casual in Somerset, at Weston-in-Gordano (9), (Sole, 1791), and at Cheddon Fitzpaine (3), 1920 (Miss A. G. Miller).

Trifolium incarnatum L. Crimson Clover

Formerly cultivated and sometimes found as an impermanent escape. Only once recorded in recent years, at Hinton St George (4), 1966 (R & IR).

Trifolium arvense L. Haresfoot Clover

Quite common in sandy places along the coast from Porlock (2) to Sand Bay (9). Rare inland and usually only a casual, perhaps introduced with sand for building work. There are recent records from districts 4, 8 and 10.

Trifolium striatum L. Knotted Clover

In dry grassy places, frequent near the coast and on Mendip. Also at Goblin Combe (9) but not seen recently on the Carboniferous Limestone near Clevedon (9). Very rare elsewhere, the only recent records being near Trull (3), 1974 (JVC), Westhay (8), 1969 (R & IR), and Shepton Montague (8), 1966 (CAH). Formerly recorded also in district 4.

Trifolium scabrum L. Rough Clover

In similar habitats and often the same places as *T. striatum.* Besides coastal localities between Minehead (2) and Sand Point (9) and others on Mendip (9) it has long been known inland at Ham Hill (4) and near Stawell (5). Other recent inland records are at Compton Dundon (5), 1968 (JGK), near Shepton Montague (8), 1966 (CAH), and on Solsbury near Bath (10), 1961 (R & IR).

Trifolium subterraneum L. Subterranean Clover

Dry grassland in sandy and gravelly places. Rare and now only known between Porlock and Minehead (2), at Cannington (3) and near Keynsham (10), though formerly recorded more widely in districts 3 and 9.

†**Trifolium glomeratum** L. Clustered Clover

Formerly near the sea at Bossington and Dunster (2) but last seen there in 1916 (NGH). Also recorded near West Monkton (3) by Marshall (1914) but not reported there since.

Trifolium suffocatum L. Suffocated Clover

Very rare and only on Minehead Warren (2) where it was first found by Marshall about 1911. There are old records from Lilstock (2) (J. C. Collins) and from Weston-super-Mare (9) (T. B. Flower) where the site was destroyed by building.

Trifolium hybridum L. Alsike Clover

An escape from cultivation naturalized by roadsides and in waste places. Common over much of the county except on Exmoor where it is rare.

Trifolium repens L. White Clover

Abundant in grassland, on roadsides and in waste places throughout the county.

Trifolium fragiferum L. Strawberry Clover

Common on droves, roadsides and in damp meadows on the Liassic and alluvial clays of central and south Somerset, and frequent on heavy soils in the north. Very rare west of Taunton (3) except near the coast between Minehead and Steart (2). Not recorded in district 1.

Trifolium resupinatum L. Reversed Clover

An alien casual imported with foreign grain, first recorded in the county by White at Portishead (9) in 1904 and later found in a few other waste places near Bristol and Bath (10). White stated it did not ripen seed in this country but it has proved remarkably persistent at Combwich (3) where it was first recorded in 1929 by G. Watts and is still firmly established.

Trifolium campestre Schreb. Hop Trefoil

In grassy and waste places, very common on calcareous and sandy soils. Less so on other soils but recorded in all districts.

Trifolium dubium Sibth. Lesser Trefoil

Very common in grassy and waste places, by roadsides and on wall tops throughout the county.

LEGUMINOSAE

Trifolium micranthum Viv. Slender Trefoil

T. filiforme L.

In thin turf on sandy or gravelly ground. More frequent near the coast between Porlock and Steart (2) than elsewhere where it is rare or overlooked. Small forms of *T. dubium* are sometimes reported as this in error but it has been reliably recorded at Cannington (3), Ham Hill (4), Brean Down and a few spots on Mendip (9), Keynsham and Mells (10).

Trifolium lappaceum L.

A rare casual found on a tip at Bedminster (10), 1978 (ALG, CML & RMP conf. CJC).

ANTHYLLIS L.

Anthyllis vulneraria L. Kidney Vetch

Rather common but local in dry grassland on sand dunes and on calcareous soils in districts 2, 8, 9 and 10. Elsewhere only south of Crewkerne (4), on the Lias near Somerton (5) and on the chalk of the eastern border (7). Not in district 1 nor in 3 apart from Miss F. Elworthy's old record "near Wellington." This species is very variable over its European range and has been divided into many subspecies; only ssp. **vulneraria,** the most frequent British form, has been identified in Somerset.

LOTUS L.

Lotus corniculatus L. Common Birdsfoot Trefoil

Very common in dry grassland and on roadsides throughout the county.

Lotus tenuis Waldst. & Kit. ex Willd. Narrow-leaved Birdsfoot Trefoil

In rough grassland on clay. Very rare and apparently decreasing. The only recent records are near Somerton (5) where it has been known for many years, on the Polden Hills near Street (8), 1974 (Mrs. F. Rhodes), at North Wootton (8), 1966 (DHP), and by Blagdon Lake (9), 1975 (JA), though formerly it was more widespread and recorded in a few places in districts 3, 4 and 10, as well as others in 5, 8 and 9.

Lotus uliginosus Schkuhr — Greater Birdsfoot Trefoil

L. pilosus sensu Murray

Damp fields, ditchsides and marshes. Common throughout the county but less so on calcareous soils than on acid ones.

Galega L.

Galega officinalis L. — Goat's Rue

Garden escape.

Robinia L.

Robinia pseudacacia L. — False Acacia

Frequently planted and spreads by suckers.

Astragalus L.

Astragalus glycyphyllos L. — Wild Liquorice

In rough pastures and on grassy lane verges. Quite frequent on the Lias in district 5 between Aller and Charlton Mackrell and at Walton. Very rare elsewhere and only seen recently near Weston-in-Gordano (9), Publow, Midford, Hinton Charterhouse and Frome (10) although formerly more widespread in the north of the county.

†**Astragalus odoratus** Lam. — Lesser Milk-vetch

A rare casual established on the towpath by the river at Bath for at least forty years but not recorded since 1963; it is feared it has been eradicated by "Tidying up" operations.

Ornithopus L.

Ornithopus perpusillus L. — Birdsfoot

In gravelly and sandy places. Frequent on Exmoor and the Brendon and Quantock Hills. Elsewhere it has only been seen recently between Chiselborough and Yeovil (4), on Pennant Sandstone near Keynsham (10) and on the Greensand near Witham (10) and Yarnfield (10) (Wilts). Previously recorded at Burnham (8), Uphill (9) and between Clevedon (9) and Long Ashton (10) but not seen in any of these places since 1939.

LEGUMINOSAE

Coronilla L.

Coronilla varia L. Crown Vetch

An uncommon casual in waste places and by railways, which has persisted since at least 1906 on walls of the Bishop's Palace at Wells (8).

Coronilla scorpioides (L.) Koch

A rare casual which has been recorded as a garden weed.

Hippocrepis L.

Hippocrepis comosa L. Horseshoe Vetch

Frequent but local in the turf of calcareous downs. On the Lias of the Polden Hills from Charlton Mackrell to Walton (5), on the Carboniferous Limestone of Mendip (8 & 9), Worle Hill, Clevedon (9) and Leigh Down (10), on the chalk near Kilmington (7) (Wilts), and on the Oolite from Bruton (8) to Bath (10). The only recent record in V.C.5 is at Cadbury Castle (5) though it was formerly recorded at other places.

Scorpiurus L.

Scorpiurus muricatus L.

Bird-seed alien.

Onobrychis Mill.

Onobrychis viciifolia Scop. Sainfoin

An escape from cultivation which becomes naturalized on calcareous soils. Quite frequent on the Lias near Somerton (5) and on the Oolite round Bath (10) but rather rare now elsewhere; recent records are near Bruton, Pilton (8), Burrington, Clevedon, Wraxall (9), Dundry and Frome (10). Not seen recently in V.C.5 although there are old records in districts 2, 3 and 4. Although both Murray and White thought it might be a native in Somerset this seems unlikely as it has not persisted where there are old records and present localities do not have a long history. Babington (1834) only gives one locality near Bath.

CICER L.

Cicer arietinum L.

An uncommon casual first recorded in the county at Brislington (10), 1916 (Misses Cobbe), but only seen recently at the Royal Portbury Dock (10), 1979 (ALG).

VICIA L.

Vicia hirsuta (L.) Gray — Hairy Tare

Common in hedges, on waste ground and particularly on railway banks throughout the county except on the higher parts of Exmoor.

Vicia tetrasperma (L.) Schreb. — Smooth Tare

V. gemella Crantz

On hedgebanks, roadsides and railway banks, rough grassland and on waste ground. Common in districts 3 and 4 and frequent elsewhere but avoiding high ground and the Levels. Not recorded recently in district 7.

Vicia tenuissima (Bieb.) Schinz & Thell. — Slender Tare

V. gracilis Lois.

In cornfields and on hedgebanks. Now very rare except in a small area on the Lias stretching from Curry Rivel (4), through Somerton (5) to Lydford (8) and north to Walton (8). Outside this area it has only been seen recently at Stogursey (2), 1968 (HWB), and Staple Fitzpaine (4), 1975 (JVC), both spots on Lias. Formerly it was more widespread in districts 2, 3 and 10. The first British record for this species was made on Barrow Hill near Bath (10) by Babington (1839) but it has not been recorded in Somerset north of Mendip since 1907.

Vicia cracca L. — Tufted Vetch

Very common in hedges, on field borders and in other grassy places over most of the county but less so in the extreme west.

Vicia tenuifolia Roth — Fine-leaved Vetch

A rare casual which sometimes persists for many years, as it has at Bruton (8) and Combe Down near Bath (10).

LEGUMINOSAE

Vicia villosa Roth — Fodder Vetch

A rare casual only seen recently on tips at Brislington (10), 1978 (ALG), and Bedminster (10), 1979 (ALG).

Vicia orobus DC. — Wood Bitter-vetch

In rough hill pasture on Mendip, now very rare, in district 9. The best-known old locality, a field on Tyning's Farm above Cheddar, was ploughed up about 1954 and the plant exterminated there. It has probably suffered a similar fate in other old localities.

Vicia sylvatica L. — Wood Vetch

On wood borders, in rides and clearings, and in thickets in rocky places. Locally common on the wooded coast near Culbone and Minehead (2) and northwards from Wells (8) and the Mendips to the area near Bristol and Bath (9 & 10). Rare elsewhere and recorded recently only at Langridge (2), Montacute (4) and Great Breach Wood (5).

Vicia sepium L. — Bush Vetch

Very common on hedgebanks, roadsides and wood borders throughout the county except in the extreme west of Exmoor.

† **Vicia lutea** L. — Yellow Vetch

Known on Glastonbury Tor (8) for over 200 years it became extinct about 1860. The earliest specimen (in the British Museum herbarium) is dated 1739. This species is considered native in Britain only near the sea and this may well have been another example of the survival of a submaritime species inland long after the sea had retreated as in the case of *Trifolium squamosum.* Elsewhere in North Somerset it has occurred in a few places as a casual but not since 1941. Davis's pre-1802 record at Midford (10), queried by Murray, was probably one of these casual occurrences.

† **Vicia hybrida** L. — Hairy Yellow Vetch

This also grew on Glastonbury Tor (8) from before 1670 (Ray) until at least 1832 (T. Clark in **TTN**). It has since occurred as a casual near Bristol (9 & 10) most recently at St Anne's (10), 1919 (CIS).

Vicia sativa L. — Common Vetch

Very common as an aggregate on roadsides, in dry grassland and in waste places throughout the county. All plants seen recently can be assigned to ssp. **nigra** (L.) Ehrh., which includes the forms usually known as *V. angustifolia* L. and *V. segetalis* Thuill. (*V. angustifolia* var. *segetalis* (Thuill.) Koch). The former, characterized by concolorous flowers (usually bright pink) and narrow upper leaflets, is common on dry sandy or gravelly soils and sometimes in poor grassland. *V. segetalis* is even commoner, particularly on calcareous soils, and is more ubiquitous, occurring in hedges and on grassy roadsides as well as in the habitats of *V. angustifolia.* The plant formerly cultivated widely as a fodder crop and occasionally found as an escape was ssp. **sativa** but it has not been identified in the county for many years. Plants recorded as *V. sativa* (s.s.) are usually robust specimens of *V. segetalis.*

Vicia lathyroides L. Spring Vetch

Rare in sandy places near the sea. Only recorded recently near Minehead (2) on North Hill towards Selworthy and on the Warren, and at Berrow, Brean Down and Kewstoke in district 9. Formerly also at Steart (2), Burnham (8), above Charlcombe Bay (9), and on Steep Holm but not seen in any of these places for many years.

Vicia bithynica (L.) L. Bithynian Vetch

In grassy lanes, on hedgebanks and in rough pastures. Rare and decreasing. Only recorded recently near Minehead and West Quantoxhead (2), Corfe (3), Compton Dundon (5), Publow and Kilmersdon (10), 1980 (DG), though formerly more widely in districts 3, 8, 9 and 10.

Vicia narbonensis L.

A rare casual.

Vicia faba L. Broad Bean

Occasionally found on rubbish tips but not persisting.

Vicia pannonica Crantz

The type of this alien species (with yellow flowers) is well established in Kent and was perhaps introduced with grass seed; a patch was found for the first time in the county at Saltford (10) in 1976 (RSC) and may persist there. The purple-flowered ssp. **striata** (Bieb.) Nyman had previously been found as a casual at Ashton Gate (10), 1922 (CIS).

LEGUMINOSAE

LATHYRUS L.

Lathyrus aphaca L. Yellow Vetchling

This species was probably introduced originally from southern Europe with imported grain and the earliest record of it in Somerset is as a cornfield weed at Claverton (10) (Davis, 1802). It has also been introduced with clover and lucerne seed. It is now rare and most recent records are from grassy lanes and hedgebanks on the Lias of districts 5 and 8 between Charlton Mackrell and Pawlett. Outside this area, where it seems firmly established, it has only been recorded recently at Thornfalcon (3), 1972 (CAH & JVC), and Farleigh Hungerford (10), 1962 (R & IR), though formerly it was more widespread and recorded in districts 2, 4, 8 and 9 also.

Lathyrus nissolia L. Grass Vetchling

On hedgebanks, in grassy lanes, on railway banks and in scrub, usually on clay. Local and perhaps under-recorded as it is hard to see except when in flower. Fairly frequent in district 2 between Minehead and Steart, in districts 3, 4, 5 and 8 between Wellington and Castle Cary and in district 10 in the Chew valley near Pensford and by the Avon estuary below Bristol. Rare outside these areas and only recorded in a few spots including Sutton Bingham (4), 1966 (JGK), North Wootton (8), 1966 (DHP), Pylle (8), 1967 (CAH), Churchill (9), 1972 (RGR), and Trudoxhill (10), 1971 (PT).

Lathyrus hirsutus L. Hairy Vetchling

A rare casual, only seen recently in the county on a rubbish tip at Street (8), 1971 (JGK). At one time this alien was considered a possible native in Britain and Somerset was included in its distribution due to an error by the Rev G. Swayne which first appeared in Withering (1796) and was repeated in many botanical books including Bentham & Hooker (ed. 8,1904) although as long ago as 1866 Jenyns had pointed out that the plant concerned was really *Vicia bithynica.*

Lathyrus pratensis L. Meadow Vetchling

Very common in hedges and grassland throughout the county.

Lathyrus tuberosus L. Tuberous Pea

An alien which has sometimes proved very persistent in rough grassland. It was known at Cheddar (9) and near Keynsham (10) for over twenty years but has apparently gone from both localities. The only recent record is near Kelston (10), 1973 (PJMN). Only once recorded in V.C.5, as a casual near Watchet (2), 1943 (R. A. Priske & WW).

Lathyrus sylvestris L. Narrow-leaved Everlasting Pea

In thickets and hedges. Widespread but not common. Recorded from all districts except 1 and 6 and not recently in 7. Absent from Exmoor.

Lathyrus latifolius L. Broad-leaved Everlasting Pea

A frequent garden escape often established in hedges and on railway banks.

Lathyrus palustris L. Marsh Pea

Very rare and now only in one marshy enclosure near Catcott (8), though formerly more widespread in the surrounding fenland. The species figured in the list of Bath plants provided by Davis for Warner (1802) and the record was copied by Babington and denounced as an error by Jenyns, Murray and White but all these botanists failed to notice that the entry had been deleted by Davis himself in the Corrigenda at the end of the list! A record for Portishead Point (9) by Dr A. Broughton (1779) was almost certainly an error. The var. **pilosus** (Cham.) Ledeb., an alien form, occurred on the dunes at Berrow (9) in 1958 (D. Munro Smith det. NYS) but did not persist there.

Lathyrus montanus Bernh. Bitter Vetch

In woods, hilly grassland and under hedges, particularly on acid soils. Frequent in the west and south, on the Greensand of the eastern border and on Mendip as far west as Shipham (9). Also on the Coal Measures and Old Red Sandstone in the north. Absent from a large area in the centre of the county including the whole of district 5, the Polden Hills and the Levels.

Lathyrus grandiflorus Sibth. & Sm.

A garden escape naturalized in one or two places near Bristol and Bath (10).

ROSACEAE

Spiraea L.

Spiraea salicifolia L. Bridewort

An occasional garden escape.

Filipendula Mill.

Filipendula vulgaris Moench Dropwort

Spiraea filipendula L.

In limestone grassland. Almost confined to the Carboniferous Limestone of district 9. Locally plentiful on Mendip from Uphill to Binegar (10), at Weston-super-Mare and Worle, Sand Point and St Thomas Head, and near Clevedon. It is also quite plentiful on the Oolite at Lansdown (10) where it was first recorded by E. Simms (Babington, 1834). Not now on Brean Down but recorded there by J. C. Melvill (Murray, 1896). Unknown elsewhere except as a casual or garden throw-out.

Filipendula ulmaria (L.) Maxim. Meadowsweet

Spiraea ulmaria L.

Very common in damp meadows, marshes and by ditches throughout the county.

Rubus L.

†**Rubus saxatilis** L. Stone Bramble

This northern species was found in two places in the same year, 1883; at Banwell Castle (9) by H. S. Thompson and in Asham Wood (10) by Murray. It has not been seen in either locality since 1920.

Rubus idaeus L. Raspberry

In woods, scrub and by roadsides. Common on acid soils and frequent elsewhere throughout the county although only an escape from cultivation in many of its localities.

Rubus phoenicolasius Maxim. Japanese Wineberry

An uncommon escape from cultivation first recorded in the county in 1957 at Chantry (10) (EDO) and seen since then in three other places in districts 9 and 10.

Rubus caesius L. Dewberry

In hedges and by roadsides. Common in most of the county, especially on calcareous soils, but rather scarce in the extreme west.

Rubus fruticosus L.sensu lato Blackberry

Although the brambles of Somerset were extensively studied by Murray, White and Marshall ideas on the identity and limits of the many microspecies of this difficult genus have been much revised in recent years. There has been no resident expert of late and much work remains to be done. The following account is based on recent records by A. Newton with additions by L. J. Margetts and C. M. Lovatt. Figures are those of 10 km. squares of the National Grid in which the taxon has been recorded: those in italics being herbarium records recently re-determined by A. Newton.

Sect. **Suberecti** P. J. Muell.

Rubus nessensis W. Hall 92, 66, 73
Rubus scissus W. C. R. Wats. 73
Rubus sulcatus Vest ex Tratt. *66*
Rubus affinis Weihe & Nees 43, 44
Rubus nobilissimus (W. C. R. Wats.) Pearsall 43, 44, *66*
Rubus divaricatus P. J. Muell. 43

Sect. **Triviales** P. J. Muell. (*Rubi Corylifolii*)

Rubus sublustris Lees 30, 57. Probably in many other squares also.
Rubus tuberculatus Bab. 57, *64*, 66

Sect. **Sylvatici** P. J. Muell.

Rubus sciocharis (Sudre) W. C. R. Wats. 73
Rubus nemoralis P. J. Muell. (*R. selmeri* Lindeb.) 11, 21, 30, 43, 44, 73
Rubus questieri Muell. & Lefèv. *73*
Rubus laciniatus Willd. A cultivated species of unknown origin occasionally found as an escape. It has persisted on Blackdown (Mendip) for at least 30 years.
Rubus lindleianus Lees *57*, 66
Rubus amplificatus Lees 30
Rubus pyramidalis Kalt. 83
Rubus albionis W. C. R. Wats. 94, 21, 30
Rubus riddelsdellii Rilstone 94,
Rubus polyanthemus Lindeb. 84, 21, 30, 43, 44, 66

ROSACEAE

Rubus rubritinctus W. C. R. Wats. Frequent in the west (districts 1 & 2). Also recorded in 30, 57 & 66

Rubus prolongatus Boulay & Letendre. Frequent in districts 1 & 2. Also recorded in 11, 21 & 66

Rubus cardiophyllus Muell. & Lefèv. 94, 21, 30, 44, *57,* 66, 73

Rubus sprengelii Weihe 44, 73

Sect. **Discolores** P. J. Muell.

Rubus ulmifolius Schott. The commonest bramble in hedges over most of the county, particularly on calcareous soils. Less frequent on the acid soils of the west and south. The only English bramble species with normal sexuality, it hybridizes with *R. caesius* and with *R. vestitus* and probably with other microspecies.

Rubus discolor Weihe & Nees (*R. procerus* P. J. Muell.). A cultivated variety (the so-called Himalayan blackberry) is a frequent escape. Recorded in districts 4, 5, 7, 8, 9 & 10

Sect. **Appendiculati** (Genev.) Sudre

Rubus surrejanus Barton & Riddelsd. (*R. lasiostachys* Muell. & Lefèv.) 73

Rubus lanaticaulis Edees & Newton (*R. hebecaulis* auct.) 57

Rubus vestitus Weihe & Nees 83, 92, 94, 57, 63, 66, 73. Probably more frequent in woodland in the east and north of the county than these records indicate.

Rubus adscitus Genev. 92, 94, 30

Rubus mucronatiformis Sudre 21, 30

Rubus drejeri Jansen *44*

Rubus leyanus Rogers 83, 94, 11, 21

Rubus dentatifolius (Briggs) W. C. R. Wats. *92, 73*

Rubus echinatus Lindl. (*R. discerptus* P. J. Muell.) 30, *57*, *66*. Probably more frequent than these records indicate.

Rubus micans Gren. & Godr. *30*

Rubus flexuosus Muell. & Lefèv. (*R. foliosus* Weihe & Nees) *73*

Rubus bloxamii Lees 11, 21, 73

Rubus fuscus Weihe & Nees 73

Rubus fuscicaulis E. S. Edees 57

Rubus pallidus Weihe & Nees *64*

Rubus glareosus Rogers (*R. argutifolius* Muell. & Lefèv.) 73

Rubus longithyrsiger Lees ex Bak. 11

Rubus rufescens Muell. & Lefèv. *36,* 57, *67*

Rubus anglosaxonicus Gelert (*R. apiculatus* Weihe & Nees) 21

Rubus heterobelus Sudre 73

Rubus raduloides (Rogers) Sudre (*R. anglosaxonicus* auct.) *30, 36,* 57, *66*

Rubus vectensis W. C. R. Wats. (*R. retrodentatus* Muell. & Lefèv.) SS73, 82, 92, 94; ST21, 30, 66, 73

Rubus leightonii Lees ex Leighton 21

Rubus diversus W. C. R. Wats. 46, 57, 66

Rubus hiernii Riddelsd. (*R. rotundifolius* (Bab.) Bloxam) *30*

Rubus hylocharis W. C. R. Wats. 94

Rubus scabripes Genev. 94, 21, 30

Rubus milesii A. Newton (*R. adenolobus* W. C. R. Wats. nom inedit.) *57*, 73

Rubus dasyphyllus (Rogers) E. S. Marshall 94, 21, 57, 66, 73

Sect. **Glandulosi** P. J. Muell.

Rubus durotrigum R. P. Murr. 73

POTENTILLA L.

Potentilla palustris (L.) Scop. Marsh Cinquefoil

Comarum palustre L.

Bogs and marshes. Rare and only on the peat moors (8) between Catcott and Glastonbury. Rutter's old record (1829) "near Axbridge" (9) has not been confirmed since and two other old records from ponds at Ashton Manor and Leigh Woods (10) were presumably introductions.

Potentilla sterilis (L.) Garcke Barren Strawberry

P. fragariastrum Pers.

Woods, hedgebanks and on walls. Very common throughout the county except on the Levels where it is scarce.

Potentilla anserina L. Silverweed

Very common on roadsides, in waste places and in poor grassland.

Potentilla argentea L. Hoary Cinquefoil

A rare but persistent casual in Somerset, at present known only on an old spoil heap at Charterhouse-on-Mendip (9) where it was first seen in 1927 by F. Langford. It was found on a rocky limestone slope near Tickenham (9) in 1915 by the Rev E. Ellman, and in another spot about a mile away in

1920 by Mrs C. I. & N. Y. Sandwith who thought it might be native but it has not been seen in the area since 1941. Other former records, at Bridgwater (5), Portishead (9) and Englishcombe (10) were more obviously casual.

Potentilla recta L. Sulphur Cinquefoil

An occasional garden escape.

Potentilla norvegica L. Ternate-leaved Cinquefoil

An uncommon casual.

Potentilla tabernaemontani Aschers. Spring Cinquefoil

P. verna L.

Very local on outcrops of the Carboniferous Limestone. On Mendip from Bleadon and Hutton to Rookham (9), in several places on Broadfield Down (9), on the ridge between Wraxall (9) and Leigh Woods (10) and in the Avon Gorge (10).
Recorded on Brean Down (9) by St Brody (1856) but not reported there since White (1912). The statement "frequent round Bridgwater" attributed to Collins by Watson (1837) must have been a mistake.

Potentilla erecta (L.) Räusch. Tormentil

P. silvestris Neck.

On heaths, moors, hedgebanks and dry pastures on acid or neutral soils. Very common over most of the county except for the Levels, where it only occurs on the peat moors (8), and on the Bath Oolite (10) where it is rare.

Potentilla anglica Laichard. Trailing Tormentil

P. procumbens Sibth.

On hedgebanks, in open woodland and on heaths. Much less common than *P. erecta* and usually in damper and shadier places. Widespread, but not recorded in districts 5 or 7 nor recently in 6 where, however, it is likely to survive and may have been overlooked. At one time this species was considered a subspecies or variety of *P. erecta* but it is now believed to have arisen by chromosome doubling (allopolyploidy) from a hybrid between *P. erecta* and *P. reptans.*

× **erecta** (*P.* × *suberecta* Zimm.). The parents hybridize freely and hybrids may be expected wherever they grow together, as on the peat moors (8).

Potentilla reptans L. Creeping Cinquefoil

Very common on roadsides, in pastures and as a weed in cultivated and waste land throughout the county.

× **anglica** (*P.* × *mixta* Nolte ex Reichb.)
× **erecta** (*P.* × *italica* Lehm.)
Hybrids resembling *P. reptans* do occur occasionally and have been recorded in the past in Somerset. It is not possible to determine the other parent although attempts have been made to do so.

DUCHESNEA Sm.

Duchesnea indica (Andr.) Focke Yellow-flowered Strawberry

Fragaria indica Andr.

A garden escape naturalized in a wood border at Long Ashton (10), 1972 (CHC), previously recorded also at Clapton (9) in 1935 but it does not seem to have persisted there.

FRAGARIA L.

Fragaria vesca L. Wild Strawberry

Very common in woods and scrub and on hedgebanks throughout the county.

Fragaria × ananassa Duchesne Garden Strawberry

A garden escape sometimes naturalized on railway banks.

GEUM L.

Geum urbanum L. Wood Avens

In woods, on hedgebanks and in waste places. Very common in all districts.

Geum rivale L. Water Avens

In damp woods and by streams. Rare and decreasing. Formerly widespread in North Somerset and still fairly frequent at the eastern end of Mendip in the area between Bruton (8), Shepton Mallet (8) and Frome

(10). The only recent records outside this area are at Brompton Regis (1), 1969 (KGM), Winsford (1), 1969 (CAH), Stawley (3), 1974 (HWB), Whitestaunton (6), 1967 (NGC), Kilmington (7) (Wilts), 1967 (CAH), Portbury (9), 1972 (AFD), and West Harptree (10), 1970 (EJL). Formerly recorded also in district 4.

× **urbanum** (*G* × *intermedium* Ehrh.). Near the parents but often in less damp parts of woods than *G. rivale.* Rare and only recorded recently near Dulverton (1), 1975 (CJG), at Whitestaunton (6), 1971 (NGC) and in district 10 at the eastern end of Mendip between Nettlebridge and Gare Hill. Formerly recorded also in district 8.

AGRIMONIA L.

Agrimonia eupatoria L. Agrimony

Hedgebanks, rough grassland, roadsides and waste places. Very common throughout the county except on Exmoor and the Brendons. Recorded in all districts but rare in 1.

Agrimonia procera Wallr. Fragrant Agrimony

A. odorata auct., non Mill.

Wood borders. Rare. The only locality in the county where it has a long history is at Asham Wood (10) where it has been known since 1917 (R. V. Sherring). The only other recent records are near Culbone (2), 1970 (CAH), Langford Budville (3), 1967 (CLH), Ham Hill (4), 1978 (CJC), Brewham (8), 1961 (PFH), Portbury (9), 1972 (AW), and Marston (10), 1961 (EDO). It was seen at Portbury in 1924 but otherwise these are all new records. In a dozen other localities where it has been recorded this century it has since apparently disappeared.

ALCHEMILLA L.

Alchemilla conjuncta Bab.

Garden escape.

Alchemilla filicaulis Buser ssp. **vestita** (Buser) M. E. Bradshaw
Lady's Mantle

A. vestita (Buser) Raunk.; *A. vulgaris* L. var. *filicaulis* Buser; *A. minor* auct.

In grassland, particularly on calcareous soils. Frequent on Mendip and in the north of the county. Rather rare elsewhere and not recorded in districts 5 or 7.

Alchemilla xanthochlora Rothm.

A. pratensis auct.

In damp pastures and by grassy roadsides. Rare and only in the west on Exmoor (1 & 2). Other records near Wellington (3) have not been confirmed as this species; *A. filicaulis* ssp. *vestita* certainly occurs in the area.

†**Alchemilla glabra** Neygenf.

There is a specimen of this northern species in the herbarium at Edinburgh marked as collected "near Bath" in 1837 but nothing further is known as to where it might have been found.

Alchemilla mollis (Buser) Rothm.

Garden escape.

APHANES L.

Aphanes arvensis L. Parsley Piert

Alchemilla arvensis (L.) Scop.

In rocky places where the soil is thin, on wall tops and a weed in cultivated ground and waste places. Common and recorded in all districts but rather sparse on the Levels.

Aphanes microcarpa (Boiss. & Reut.) Rothm. Slender Parsley Piert

In bare places, cultivated land and wall tops, with a decided preference for acid soils. Common on Exmoor and frequent on the Greensand of the eastern border. Rare or overlooked elsewhere. Not recorded in district 9. Only recognized as a British plant since 1949 and its detailed distribution in the county is still imperfectly known.

SANGUISORBA L.

Sanguisorba officinalis L. Great Burnet

Very local in damp meadows in the Exe and Barle valleys in district 1. Apparently only a casual elsewhere. Recorded near Washford and Williton (2) in 1928, near West Camel (5) in 1878, and near Wells (8) in 1886 but not seen again in any of these places. Recent reports from railway banks in districts 9 and 10 could not be refound and verified.

ROSACEAE

Sanguisorba minor Scop. ssp. **minor** Salad Burnet

Poterium sanguisorba L.

In dry grassland on calcareous soils. Common over much of the county except for the Levels and the acid soil areas. Absent from Exmoor and not recorded anywhere in district 1.

ssp. **muricata** Briq. Fodder Burnet

Poterium polygamum Waldst. & Kit.

Formerly cultivated as fodder and occasionally found as an escape. Recorded recently only at Pylle (8), 1966 (CAH), and Belmont Hill, Wraxall (9), 1974 (CHC).

Rosa L.

Rosa arvensis Huds. Field Rose

In hedges and rough fields throughout the county. Hybrids of this species with both *R. stylosa* and *R. canina* have been recorded in V.Cs. 5 and 6.

Rosa pimpinellifolia L. Burnet Rose

R. spinosissima L. pro parte

On sandhills and cliffs, usually near the sea. Very rare and perhaps extinct as a native plant though sometimes found as a garden escape. Formerly recorded on the coast near Watchet (2), where it was last seen in 1956, and near Weston-super-Mare, Kewstoke and Clevedon (9). In 1963 and 1964 a number of bushes were observed on high cliffs of the Cheddar Gorge (9) (PJMN & AJW), probably naturalized escapes as the species had not been previously recorded in the area.

Rosa rugosa Thunb.
Rosa virginiana J. Herrmann

Occasional garden escapes.

Rosa stylosa Desv.

Common and widespread, particularly on calcareous soils. The most frequent form is var. **systyla** (Bast.) Bak. Recorded in all districts but scarce in the extreme west.

Rosa canina L. Dog Rose

Very common in hedges throughout the county. Many varieties have been named and some have been distinguished as separate species including **R. dumalis** Bechst. and **R. corymbifera** Borkh. (*R. dumetorum* Thuill.), both of which are frequent in Somerset. The distribution of the other varieties has not been studied recently. Hybrids with *R. stylosa* have been recorded in both V.Cs 5 and 6.

Rosa obtusifolia Desv.

R. borreri Woods

An uncommon rose of the *Canina* group which has been recorded in the past in all districts except 5 and 7 but which has been overlooked of recent years because of identification difficulties.

Rosa tomentosa Sm. Downy Rose

In hedges and rough hilly pastures. Rather rare. Most recent records are from the eastern half of the county in districts 5, 8 and 10.

Rosa sherardii Davies

Hedges. Rare and mainly on acid soils in the west and south. Specimens determined by R. Melville from districts 1 and 4 were mostly forma **resinosoides** (Crép.) W.-Dod but var. **omissa** (Déségl.) W.-Dod also occurs.

Rosa rubiginosa L. Sweet Briar

Bushy hillsides on calcareous soils. Reliably recorded recently only on the Lias of district 5 near Somerton and on Tickenham Hill (9), 1962 (RMB).

Rosa micrantha Borrer ex Sm.

Formerly widespread in hedges and bushy places but not recorded recently except at Cleeve Hill (2) and Cheddar (9); probably overlooked.

†**Rosa agrestis** Savi

Formerly in bushy places near the sea in districts 2 and 9 but now apparently extinct. Surviving bushes on St Thomas Head (9) were examined in 1954 by R. Melville who considered most of them to be hybrids between *R. agrestis* and *R. stylosa*. A single bush found by Marshall between

Somerton and Compton Dundon (5) was identified as this species by A. H. Wolley-Dod.

PRUNUS L.

Prunus spinosa L. Blackthorn; Sloe

In hedges and woods. Very common throughout the county.

Prunus domestica L. Wild Plum

The ssp. **institia** (L.) C. K. Schneid. (Bullace) is widespread in hedges and wood borders, particularly in north Somerset. The cultivated ssp. **domestica** sometimes occurs as an escape. Plants intermediate between *P. spinosa* and *P. domestica* are frequent and are probably hybrids between them.

Prunus cerasifera Ehrh. Cherry Plum

Occasionally planted in hedges.

Prunus avium L. Wild Cherry

In woods and hedges. Common except in the lowlands and recorded from all districts.

Prunus cerasus L. Dwarf Cherry

In hedges on acid soils. Only seen recently in the west of the county.

Prunus padus L. Bird Cherry

Only a planted tree in Somerset.

Prunus laurocerasus L. Cherry Laurel

Originally introduced but now naturalized and widespread in woods, plantations, parks and on railway banks.

Prunus lusitanica L. Portugal Laurel

Often planted in ornamental grounds but not naturalized.

COTONEASTER Medic.

Cotoneaster simonsii Bak. Himalayan Cotoneaster

Planted in gardens and as an undershrub in coverts and sometimes bird-sown and found on wood borders, in hedges and in waste places. Recorded from all districts except 1, 3 and 4.

Cotoneaster horizontalis Decne. Wall Cotoneaster

Garden escape.

Cotoneaster microphyllus Wall. ex Lindl. Small-leaved Cotoneaster.

A bird-sown alien originating from gardens but now well established in many places on the Carboniferous Limestone of Mendip and other outcrops in the north (8, 9 & 10) and on the Oolite round Bath (10). Occasionally on walls elsewhere but not yet recorded in districts 1, 2, 3, 6 or 7.

Cotoneaster frigidus Wall. ex Lindl.

Garden escape.

CRATAEGUS L.

Crataegus laevigata (Poir.) DC. Midland Hawthorn

C. oxyacanthoides Thuill.; *C. oxyacantha* auct.

Very rare in Somerset which lies at the south-western extreme of its British distribution. First reported at Odd Down, Bath (10) in 1935 (DEC) and subsequently in Stockwood Lane, Whitchurch (10), 1962 (IWE conf. AJW). It is feared it has been overwhelmed by building development in both sites and at present the only locality known in the county is a hedge near Evercreech (8), 1967 (CAH). Specimens at **TTN** and **LDS** from two trees reported on Tickenham Moor (9) in 1909 by Miss M. A. G. Livett have since been determined as forms of *C. monogyna.*

Crataegus monogyna Jacq. Hawthorn; May

C. oxyacantha L. pro parte

Abundant in hedges, woods and thickets throughout the county. A pink variety of garden origin is quite often seen as a hedgerow tree. The famous Glastonbury Thorn is an early-flowering variety and several trees still flourish there. The legend of its connection with St Joseph of Arimathea is apparently relatively recent as the earliest known reference to this is as late as 1714 although the hawthorn "grene all the wynter" is referred to in Turner's Herball (1562).

ROSACEAE

Crataegus laciniata Ucria

C. orientalis Pall. ex Bieb.

A planted tree of this species not far from Shapwick station (8) has been reported by many botanists since 1921 and still survives there.

MESPILUS L.

Mespilus germanica L. Medlar

Occasionally planted and bird-sown escapes have sometimes persisted in "wild" situations for many years. It has been known in Goblin Combe (9) since 1908 and a tree on the undercliff at Clevedon (9), first seen in 1879, was last recorded there in 1938.

A hybrid with *Crataegus monogyna* (x *Crataemespilus grandiflora* (Sm.) Camus) was reported from Goblin Combe in 1964 (Dr J. E. Fraymouth & Dr L. E. Hawker).

SORBUS L.

Sorbus aucuparia L. Rowan

Pyrus aucuparia (L.) Ehrh.

Woods and hedges particularly on acid soils. Common on Exmoor, the Quantock and Blackdown Hills and in the woodland on the Wiltshire border. It also occurs on Mendip and on the hills between Clevedon (9) and the Avon Gorge (10). Widespread elsewhere as a planted tree.

Sorbus intermedia (Ehrh.) Pers. Swedish Whitebeam

A planted tree which regenerates in Leigh Woods (10) and may do so elsewhere.

Sorbus anglica Hedl.

Pyrus intermedia sensu White

An endemic British species of limestone rocks closely related to the European *S. austriaca* (G. Beck) Hedl. Very rare and only in a few spots on Mendip (9) and in Leigh Woods (10). Although first remarked on in the Cheddar Gorge in 1901 by the Rev A. Ley this species was not described and named by Hedlund until 1914.

Sorbus aria (L.) Crantz — Common Whitebeam

Pyrus aria (L.) Ehrh.

Common in woods on the limestones of north Somerset in districts 7, 8, 9 and 10. Occasionally planted elsewhere.

× **aucuparia** (*S.* × *thuringiaca* (Ilse) Fritsch; *Pyrus pinnatifida* sensu White). A rare hybrid which is known in Leigh Woods (10). "*Pyrus pinnatifida*" was recorded there by S. Rootsey in 1828 and by others at later dates but it is not certain when the tree was first correctly identified. Cultivated forms are sometimes planted.

(× **rupicola.** This rare hybrid grows with the parents in the Wye Valley and is said to have been found on the Somerset side of the Avon Gorge (10) but this is questionable; *S. rupicola* (Syme) Hedl. has never been found in the county.)

Sorbus wilmottiana E. F. Warb.

This and the next two species are British endemics related to the European *S. graeca* (Spach) Kotschy. This one is the rarest and most restricted in distribution being found only in the Avon Gorge. It was first recorded on the Somerset side (10) in 1962 (PJMN).

Sorbus eminens E. F. Warb.

Restricted to rocky woods on limestone in the Wye Valley and Avon Gorge. First identified on the Somerset side of the Avon Gorge (10) in 1942 by S. M. Walters as a species described by E. F. Warburg but not validly named by him until 1957.

Sorbus porrigentiformis E. F. Warb.

Very rare and only reliably identified in the Cheddar Gorge, Burrington Combe, on Worlebury Hill (9) and in Leigh Woods (10). Also recorded in Asham Wood (10), 1958 (PJMN), but not seen there recently. Small-leaved forms of *S. aria* have been mistaken for this. It has been suggested that records under *Pyrus rupicola* in White (1912) were of this but there is no supporting evidence from herbarium specimens.

Sorbus vexans E. F. Warb.

Pyrus rupicola sensu Marshall

Another rare British endemic which is found only in rocky woods on the north coast of Devon and in Somerset between the county border and Culbone (2). It is related to *S. rupicola* (Syme) Hedl. Unlike the other species of the *S. aria* aggregate it does not grow on limestone.

Sorbus latifolia (Lam.) Pers. sensu lato Broad-leaved Whitebeam

The species in this aggregate are believed to have originated by hybridization between *S. aria* and *S. torminalis*. The primary hybrid between these species (*S.* × *confusa* Gremli ex Rouy & Camus; *S.* × *vagensis* Wilmott) was thought to be confined in Britain to the Wye Valley until in 1968 J. F. Archibald reported it in Weston Big Wood (9). Its presence there was confirmed in 1978 when P. J. M. Nethercott located three trees.

Sorbus bristoliensis Wilmott

Pyrus latifolia sensu White

Rare and endemic to rocky woods on the limestone of the Avon Gorge. Recorded in Leigh Woods (10) by Miss Atwood (Swete, 1854) as *Pyrus aria γ intermedia* Ehrh., it was first described as a separate species by Wilmott in 1934.

Sorbus subcuneata Wilmott

Pyrus latifolia sensu Murray pro parte

A very rare endemic species of rocky woods on the coast between Watersmeet in Devon and Minehead (2).

Sorbus latifolia (Lam.) Pers. sens. strict.

A cultivated tree, introduced from France, which was found in Leigh Woods (10) in 1957 (PJMN) and has since been observed to be regenerating there.

Sorbus torminalis (L.) Crantz Wild Service Tree

Pyrus torminalis (L.) Ehrh.

In woods and hedges on clay and limestone. Rather rare and local. On the Lias on the northern slopes of the Blackdown Hills (3 & 4), on the flanks of Quantock (3), in some of the Mendip woods (9) and in Leigh Woods (10). Less frequent than formerly elsewhere in North Somerset but it has been seen recently at Evercreech (8), 1967 (CAH), Weston-in-Gordano (9), Winford and in Friary Wood (10) where it was recorded by Sole (1791).

PYRUS L.

Pyrus communis L. Wild Pear

Isolated trees recorded in field borders and hedges are probably relics of cultivation or derived from cultivated trees.

MALUS Mill.

Malus sylvestris Mill. Crab Apple

Pyrus malus L.

In woods and hedgerows. Common over much of the county except in the west where it is scarce and only on the fringes of Exmoor and other high ground. Many wild trees are ssp. **mitis** (Wallr.) Mansf. and are the descendants of the widely cultivated ones but the native plant, ssp. **sylvestris** is also widespread, especially in old woodland.

CRASSULACEAE

SEDUM L.

Sedum telephium L. Orpine

Rare and local in woods and on shady roadsides. In many places only a garden escape but probably native in the woodlands between Penselwood (7) and Witham (8), and on Mendip between Sandford (9) and Frome (10).

Sedum spurium Bieb. Caucasian Stonecrop

A garden plant which sometimes becomes naturalized on hedgebanks and stone walls.

Sedum dasyphyllum L. Thick-leaved Stonecrop

Originally introduced but now long established and remarkably persistent on old walls. Recently recorded at Kilton (2), Muchelney (4), Long Sutton, Ashcott and North Barrow (5), Chapel Allerton (8), around Nailsea, Tickenham and Wraxall (9), near Upton Noble, Cranmore, Newton St Loe and Keynsham (10). In most of these localities the records date back fifty to a hundred years.

Sedum anglicum Huds. English Stonecrop

On acid rocks, walls and in sandy places. Common in the west on Exmoor

(1 & 2) and the coast west of Minehead (2). The only recent records away from this area are on Quantock near Halsway (2), and on the Blackdown Hills near Churchstanton (6) (V.C.3). Also as a garden escape at Middlezoy (5) and Clapton-in-Gordano (10).

Sedum album L. White Stonecrop

Frequent and widespread on rocks and walls in all districts. A well-established escape in most instances but possibly native on limestone rocks at the eastern end of Mendip near Leighton (10).

Sedum acre L. Biting Stonecrop

Very common on rocks, walls and sand dunes over most of the county. Less so in the extreme west but recorded in all districts.

Sedum forsteranum Sm. Rock Stonecrop

S. rupestre L.

On cliffs and rocks. Rare and only on the coast west of Minehead (2), in the Cheddar Gorge, on Worle Hill and in Goblin Combe (9). Formerly on the Somerset side of the Avon Gorge (10) but not seen there recently. Recorded as an escape in a few other places.

Sedum reflexum L. Reflexed Stonecrop

A widespread garden escape well established on walls in all districts.

SEMPERVIVUM L.

Sempervivum tectorum L. House-leek

Sometimes planted on walls and roofs.

CRASSULA L.

Crassula tillaea L.-Garland Mossy Stonecrop

Tillaea muscosa L.

Very rare and only in bare places on tracks in a small area on the Upper Greensand between Yarnfield Gate (7) and Gare Hill (10). Part of this area is now in Wiltshire though all is in V.C.6. First detected in 1944 by J. D. Grose.

UMBILICUS DC.

Umbilicus rupestris (Salisb.) Dandy Navelwort

Cotyledon umbilicus auct.

Common on rocks, walls and hedgebanks over much of the county except the central lowlands where it is less frequent.

(**Umbilicus erectus** DC. (*Cotyledon lutea* Huds.) was said by Hudson to grow in Somerset but this must have been an error.)

SAXIFRAGACEAE

SAXIFRAGA L.

Saxifraga spathularis Brot. × **S. umbrosa** L. (*S.* × *urbium* D. A. Webb) London Pride

A cultivated hybrid which is sometimes found as a garden outcast or escape.

Saxifraga tridactylites L. Rue-leaved Saxifrage

On wall tops, rocks and in sandy places. Very common on Mendip and in other limestone areas but decidedly rare in the west and south in districts 2, 3 and 6. Not recorded recently in district 1.

Saxifraga granulata L. Meadow Saxifrage

In rough pastures, open woodland and on chalk downs. Rare and local. Recorded recently near Yeovil (4), South Cadbury (5), White Sheet Downs (7) (Wilts), Stantonbury and Lansdown near Bath (10). It was also found in Horner Woods (2) in 1937 (Miss Ellen) and still survives although probably not native there.

Saxifraga hypnoides L. Mossy Saxifrage

On limestone rocks. Very rare and only in the Cheddar Gorge (9) where it has been known since first discovered there by Dillenius in 1726. This is the southernmost British station for this mainly northern plant. Grown in gardens and an escape in one or two other localities.

PELTIPHYLLUM

Peltiphyllum peltatum (Torrey) Engler Umbrella Plant

Garden escape, naturalized at Charterhouse-on-Mendip (9), 1979 (JA).

SAXIFRAGACEAE

CHRYSOSPLENIUM L.

Chrysosplenium oppositifolium L. Opposite-leaved Golden Saxifrage

By streams, springs and in wet shady places especially on acid soils. Common over much of the county and recorded from all districts but absent from large tracts of the lowlands in districts 3, 4, 5, 8 and 9.

Chrysosplenium alternifolium L. Alternate-leaved Golden Saxifrage

Streamsides, swamps and in wet woods. Rather rare and local, and perhaps sometimes overlooked, it is recorded from all districts except 5. Absent from the lowlands and the coastal areas.

PARNASSIACEAE

PARNASSIA L.

† **Parnassia palustris** L. Grass of Parnassus

Stated by Sole in 1782 to grow "in old pits on Burtle Moor" (8), it was not included in the list of "the more rare plants growing in the county" supplied by him for Collinson's *History* (1791) and it has never been found in the county by any later botanist. The species has declined in the south of England and must have been extinct in Somerset for nearly two centuries.

HYDRANGEACEAE

PHILADELPHUS L.

Philadelphus coronarius L. Syringa

A shrub of garden origin occasionally found naturalized by railways and streams some way from houses.

ESCALLONIACEAE

ESCALLONIA Mutis ex L.f.

Escallonia macrantha Hook. & Arn.

An escaped hedging shrub naturalized on the coast near the Devon border (2).

GROSSULARIACEAE

Ribes L.

Ribes rubrum L. Red Currant

R. sylvestre (Lam.) Mert. & Koch

In woods and hedges and on river banks. Rather common except near the coast and recorded in all districts. A garden escape in many places but probably native in woodland in the eastern part of the county.

Ribes nigrum L. Black Currant

In woods and by streams and rivers. Not common but widespread and recorded in all districts except 7. Sometimes an escape from cultivation but probably native in some woods.

Ribes sanguineum Pursh Flowering Currant

Garden escape.

Ribes uva-crispa L. Gooseberry

R. grossularia L.

In woods and hedges. Common and recorded in all districts. Usually bird-sown and so commonly cultivated it is impossible to determine whether it is native in any particular locality.

DROSERACEAE

Drosera L.

Drosera rotundifolia L. Round-leaved Sundew

In peaty bogs, usually with *Sphagnum*. Common on Exmoor (1 & 2), at the north end of Quantock (2 & 3), on the Blackdown Hills (3. 4 & 6) and on the peat moors (8). Now very rare on Mendip and only seen recently on Blackdown (9) although formerly recorded also near Priddy (9) and Downhead (10). Recorded by Murray (1896) from Gasper Common (7) (Wilts) but not seen there since. It has also gone from the moors near Wiveliscombe (3) where it was recorded by Marshall (1914).

Drosera anglica Huds. Great Sundew

In wet peaty bogs. Very rare and perhaps extinct. Recorded by Sole (1791) on Blackdown and known at Britty Common (4) for many years but it was last seen there in 1948. Murray (1896) had it on Chard Common (6) but it has not been recorded there since. Specimens from near Wedmore (8) exist in the Boswell herbarium and in 1970 J. K. Hibberd, in the course of a survey of the peat moor, found a few plants with leaves exceeding 2 cm. which may have been this species.

× **rotundifolia** *(D. × obovata* Mert. & Koch). Recorded with the parents at Britty Common (4) in 1915 by Marshall but not seen there since 1922.

Drosera intermedia Hayne Long-leaved Sundew

D. longifolia auct.

In peaty bogs. Rare and only in a few spots on the Blackdown Hills and Chard Common (6) and on the peat moors (8). Formerly at Britty Common (4) (last seen 1945) and on Blackdown, Mendip (9) but not recorded there since White's day.

LYTHRACEAE

LYTHRUM L.

Lythrum salicaria L. Purple Loosestrife

By rivers, streams, ditches and in other wet places. Common in the lowlands. Absent from the higher ground particularly in the west. Not seen recently in district 1.

Lythrum hyssopifolia L. Grass Poly

A rare casual which has only twice been recorded in the county; in gardens at Baltonsborough (8), 1882 (RPM), and at Wraxall (9), 1971 (H. Kay).

Lythrum portula (L.) D. A. Webb Water Purslane

Peplis portula L.

On the margins of ponds, damp tracks and other wet places on acid soils. Fairly frequent on Exmoor and the Brendon, Quantock, Blackdown and Mendip Hills. Rare on the eastern Greensand (7 & 8), on the peat moors (8) and at Lord's Wood, Hunstrete (10). Not recorded in district 5.

Lythrum junceum Banks & Sol.

L. meonanthum Link ex Steudel

A bird-seed alien found at Brislington (10), 1920 (CIS) and 1979 (ALG).

THYMELAEACEAE

DAPHNE L.

†**Daphne mezereum** L. Mezereon

Formerly in woods on the limestone but now extinct as a wild plant. The oldest records are from woods round Bath (10) but it has long since disappeared from that area. On Mendip the best-known locality was at Churchill Batch (9) where it could not be found after 1885 until 1915 when two plants were seen by Mrs. Sandwith. It continued there until 1954 when E.F. Payne reported it had all been dug up, the culprit having been seen. Other localities were at Compton Martin Wood (9) and Eaker Hill, Chewton Mendip (10), where it was last recorded in 1926. White (1912) gives a few other records which are believed to have been bird-sown garden escapes.

Daphne laureola L. Spurge Laurel

In woods and hedges. Quite common on the limestone but rare elsewhere particularly in the west. Recorded from all districts.

× **mezereum** (*D.* × *houtteana* Lindl. & Paxt.) was found in 1907 on a wood border near Somerton (5) by Marshall. *D. laureola* is common in this area; *D. mezereum* has never been recorded there in the wild but is grown in gardens.

ELAEAGNACEAE

HIPPOPHAE L.

Hippophae rhamnoides L. Sea Buckthorn

A native of the east coast which has been planted on the coast between Minehead (2) and Sand Point (9) in several places. On the dunes between Burnham (8) and Brean (9), where it was first introduced about 1890, it has spread remarkably and now forms impenetrable thickets.

ONAGRACEAE

EPILOBIUM L.

Epilobium hirsutum L. Great Willowherb

In ditches, by rivers and streams and in other wet places. Very common throughout the county except in the west on the higher parts of Exmoor and the Brendons.

× **parviflorum** (*E.* × *intermedium* Ruhmer non Mérat)
× **montanum** (*E.* × *erroneum* Hausskn.)

Both these relatively uncommon hybrids have been recorded in the county but not recently.

Epilobium parviflorum Schreb. Hoary Willowherb

In ditches, marshes and as a weed on waste ground. Common throughout the county.

× **roseum** (*E.* × *persicinum* Reichb.). Recorded only from Westbury Combe (9), 1895 (Rev A. Ley) and St Catherine's Court (10), 1954 (G. W. Garlick).

Epilobium montanum L. Broad-leaved Willowherb

Roadsides, hedgebanks, woods and a weed in gardens and waste places. Very common in all parts of the county.

× **parviflorum** (*E.* × *limosum* Schur). Apparently one of the more frequent hybrids with records from districts 2, 3, 5, 8 and 10. The earliest is for Copley Wood, near Somerton (5), where it was collected by Marshall in 1916 (**TTN**).

× **roseum** (*E.* × *mutabile* Boiss. & Reut.). Recorded from districts 3, 5 and 10 but not seen recently.

× **obscurum** (*E.* × *aggregatum* Celak). Recorded from districts 3 and 10 but not since 1918.

Epilobium lanceolatum Seb. & Mauri Spear-leaved Willowherb

On hedgebanks, walls, quarry and mine spoil and disused railway tracks, mainly in the sandstone districts. More widespread than formerly but still rather rare and not yet recorded in districts 5 or 7, nor recently in 8.

Epilobium roseum Schreb. Pale Willowherb

In damp places, on roadsides and as a weed in gardens. Frequent in the eastern half of the county but rare in the west except on the Lias of district 2. Absent from district 1.

Epilobium adenocaulon Hausskn. American Willowherb

This North American species, first recorded in Britain in 1891, did not reach Somerset until 1936 when it was found as a casual at Portishead Dock (9) (NYS & JPMB). It began to spread about ten years later and by

1955 was apparently spreading rapidly and hybridizing with the native species. It is now common in woods as well as on roadsides and in waste places in all districts. Some authorities now consider it conspecific with **E. ciliatum** Raf., another North American species.

× **parviflorum.** Recorded only from the Gordano valley (9), on Weston Moor in 1951 (C & NS det. G. M. Ash) and on Walton Moor in 1961 (NYS).

× **montanum.** First recorded in the county at Portishead (9) in 1938 (JPMB). Probably quite frequent but overlooked. The only records during the current survey were from Nettlecombe (2) and Wellow (10).

× **obscurum.** Recorded only from Walton-in-Gordano Moor (9), 1961 (NYS).

Epilobium tetragonum L. Square-stalked Willowherb

E. adnatum Griseb.

By streams and ditches and in other damp places, and as a weed in gardens and on roadsides. Frequent and widespread in all districts.

ssp. **lamyi** (F. W. Schultz) Nyman (*E. lamyi* F. W. Schultz). Rare or overlooked as it is not easily distinguished with certainty from ssp. **tetragonum.** Recorded at one time or another from all districts except 2, 6 and 7. The only recent records are from Wells (8) and Radstock (10), 1961 (T. D. Pennington). A hybrid between the two subspecies was recorded in 1915 in Copley Wood, near Somerton (5) by Marshall.

× **parviflorum.** Recorded from Copley Wood (5), 1915 (ESM), and Stawell (5), 1940 (WW).

× **montanum.** Recorded from Taunton (3), 1940 (WW), Shapwick (8), 1932 (WW), and Corston (10), 1912 (D. Fry).

× **lanceolatum.** Recorded by D. Fry from Brislington (10) (White, 1912) but there is some doubt about the occurrence of this hybrid in Britain as the only specimens in **BM** have been determined by G. M. Ash as other species or hybrids.

× **obscurum.** Recorded from East Quantoxhead (2), 1918 (WW), Corfe (3), 1918 (WW), Copley Wood (5), 1915 (ESM), Brislington (10), 1890, and Corston (10), 1912 (D. Fry).

Epilobium obscurum Schreb. Short-fruited Willowherb

Hedgebanks, ditches and damp ground. Very common and widespread.

× **parviflorum** (*E.* × *dacicum* Borbás). Recorded from all districts except 1, 4 and 5, and most recently in 1961 from Shapwick (8) and Walton-in-Gordano (9) (NYS).

Epilobium palustre L. Marsh Willowherb

In marshes and bogs. Common on the acid soils of Exmoor and the Brendon, Quantock and Blackdown Hills and along the eastern borders. Also on the peat moors and on Mendip. Rare elsewhere and only in a few scattered places, including some in district 5, usually in stagnant moor ditches where locally acid conditions have developed.

Epilobium brunnescens (Cockayne) Raven & Englehorn New Zealand Willowherb

E. nerterioides auct.; *E. pedunculare* auct.

An introduced species from New Zealand which has spread rapidly on acid moorland in northern and western Britain. First detected in Somerset on Quantock near Over Stowey (3), 1956 (OMH), and since recorded at Broomstreet, 1958 (AD & OMH), Croydon Hill, 1960 (Miss P. P. Hill), and Hoccombe Water, 1971 (R & IR), all in district 2. An aggressive weed in the gardens of East Lambrook Manor (4), 1978 (CAH).

Epilobium angustifolium L. Rosebay Willowherb

Chamaenerion angustifolium (L.) Scop.

Wood borders, roadsides, railway banks and waste places. Although once rare in England it was described by Murray (1896) as "rather common" in Somerset. It was perhaps most plentiful on Exmoor and the Quantocks. Since then it has spread to all parts of the county and has become very common in all districts.

OENOTHERA L.

A genus from America originally introduced as garden plants. Several species are involved and there has been much confusion over their identification and nomenclature but some clarification has resulted from a visit to Britain in 1977 by Dr Rostanski from Poland who examined specimens in many important herbaria as well as visiting the dunes in South Wales.

Oenothera biennis L. Common Evening Primrose

This was apparently one of the earliest species introduced and most of the old records were listed under this name. Whether it has really declined is not certain but it is now only a rare escape in Somerset except on the dunes at Berrow (9) where it was first recorded in 1837.

× **erythrosepala.** Recorded from North Somerset in 1883.

Oenothera erythrosepala Borbás Large-flowered Evening Primrose

This species is believed to have originated in Europe from hybridization between *O. biennis* and some other introduced species. It is the largest-flowered and now the commonest species in Britain. The first published record for Somerset was dated 1952 but it must have been present long before that. Now well established on sand dunes, railway banks and in waste places in all districts except 1. Still rare in the west except near the sea in district 2.

Oenothera cambrica Rost.

O. parviflora auct.; *O. ammophila* auct.

A new species first described by Dr Rostanski in 1977 from plants growing plentifully on the dunes of South Wales. It also grows on the dunes between Burnham (8) and Brean (9). Although Dr Focke identified this plant as his *O. ammophila* (Marshall, 1914) Dr Rostanski maintains he has seen no British specimen of the latter. Some casual occurrences elsewhere in North Somerset were probably also this species; a specimen from Portishead (9) dated 1941 was definitely this.

Oenothera muricata L.

O. rubricaulis Klebahn; *O. parviflora* auct.

There has been some doubt about what species Linnaeus intended by this name but Dr Rostanski has established this from a specimen in the Linnaean herbarium. Specimens collected at Berrow (9) in 1951 by C. I. & N. Y. Sandwith, previously named as *O. parviflora* by Professor Munz, have been identified as *O. muricata* by Dr Rostanski. The only other British specimens seen by him came from S. Lancs. in 1965.

Oenothera stricta Ledeb. ex Link Fragrant Evening Primrose

O. odorata auct.

Long established on sand dunes near Minehead (2) and between Burnham (8) and Brean (9). A rare casual elsewhere.

ONAGRACEAE

Fuchsia L.

Fuchsia magellanica Lam.

Garden escape established on North Hill, Minehead (2).

Circaea L.

Circaea lutetiana L. Enchanter's Nightshade

In woods, on shady hedgebanks and a weed in gardens. Very common in all districts.

HALORAGACEAE

Myriophyllum L.

Myriophyllum verticillatum L. Whorled Water-milfoil

In stagnant ditches on the Levels and in the Kennet & Avon canal (10). Rare and local. Recent records in moor ditches are at North Petherton (3), Kingsbury Episcopi (4), Stawell (5), Glastonbury and Huntspill (8), and Weston-in-Gordano (9). Not seen recently on the peat moor (8) although frequent there formerly.

Myriophyllum spicatum L. Spiked Water-milfoil

In small rivers, canals, moor ditches and ponds. Frequent in the Bridgwater & Taunton canal (3), the River Isle and other tributaries of the Parrett (4 & 5), ditches near the coast between Brent Knoll and Portishead (9), in the River Frome and in the Kennet & Avon canal (10). Elsewhere it has been seen recently only in the River Avill at Dunster (2), reservoirs at Litton and Barrow Gurney (10) and in a few isolated ponds and ditches including some in district 8.

Myriophyllum alternifolium DC. Alternate Water-milfoil

In streams, ditches and ponds but only in acid water. Frequent on Exmoor in and by the Exe and Barle in district 1 and the streams feeding the Badgworthy and Oare Waters in district 2. Very rare elsewhere. The only recent records are from Weston-in-Gordano Moor (9) and ponds on Mendip near Priddy (9), 1973 (JA), and on Beacon Hill (10), 1971 (CAH). Formerly recorded also from the Bridgwater & Taunton Canal (3), Chard reservoir (4) and Nailsea Moor (9).

HIPPURIDACEAE

Hippuris L.

Hippuris vulgaris L. Mare's-tail

In ponds, reservoirs and stagnant moor ditches. Very rare in South Somerset being recorded only from the Sutton Bingham reservoir and near Ilminster (4), 1969 (OEC). More frequent in moor ditches in districts 5, 8 and 9. In the Blagdon (9) and Chew Valley (10) Lakes, in several ponds in districts 8, 9 and 10, including those at Prior Park, Bath, where it was recorded by Babington in 1839, and in the Kennet & Avon Canal at Bath (10).

CALLITRICHACEAE

Callitriche L.

Callitriche stagnalis Scop. Common Water-starwort

In ditches, streams, ponds and often on the muddy margins of such places. Very common throughout the county.

Callitriche platycarpa Kütz. Various-leaved Water-starwort

C. polymorpha auct.; *C. palustris* auct.; *C. vernalis* auct.

Apparently as widespread but not so common as *C. stagnalis* and sometimes in more quickly-moving water. Its distribution is imperfectly known because of difficulties in identification; it is often sterile.

Callitriche obtusangula Le Gall Blunt-fruited Water-starwort

In ditches, streams and ponds, usually in base-rich water. Common throughout the Levels in districts 3, 4, 5, 8 and 9. Rare elsewhere, and recorded recently only from Bossington (2), 1974 (CJG), North Barrow (5), 1968 (CAH), the River Axe near Winsham (6), 1976 (JGK), Chilcompton (10), 1966 (RGR), and Trudoxhill (10), 1969 (LFS).

Callitriche hamulata Kütz. ex Koch Intermediate Water-starwort

C. intermedia Hoffm.

In rivers and streams, ditches and ponds, often in acid water. Not common although records are scattered through all districts except 6 and 7.

CALLITRICHACEAE

Callitriche truncata Guss. Short-leaved Water-starwort

In streams and ditches. Very rare and only seen recently in the Rivers Isle (4), 1972 (JGK), and Axe (6), 1976 (JGK). Unknown in the county until 1907 when it was found in the Chard reservoir (4) by Marshall. Since then it has been recorded in the Bridgwater & Taunton canal (3), at Cannington (3), Stawell (5), Sharpham (8) and Kingston Seymour (9) but it has not persisted in any of these places.

LORANTHACEAE

VISCUM L.

Viscum album L. Mistletoe

Parasitic on various trees especially apple, lime, poplar and hawthorn. Only once recorded on oak, in Leigh Woods (10), 1916 (IMR). Very common where there are extensive apple orchards in districts 3, 4, 5, 8, 9 and 10. Rare outside these areas and in districts 2, 6 and 7. Absent from district 1 and the higher ground of Exmoor and the Quantock, Blackdown and Mendip Hills.

SANTALACEAE

THESIUM L.

†**Thesium humifusum** DC. Bastard Toadflax

Formerly in grassland over a limited area of Bathampton Down (10) where it had been known since 1839. In 1933 W. D. Miller wrote that it was not then in danger of extinction but it has not been recorded there since although searched for by many botanists.

CORNACEAE

CORNUS L.

Cornus sanguinea L. Dogwood

Swida sanguinea (L.) Opiz; *Thelycrania sanguinea* (L.) Fourr.

In hedges, woods and scrub particularly on calcareous soils. Very common throughout the county except on Exmoor and the Brendons where it is rare.

Cornus sericea L.

C. stolonifera Michx.; *Swida sericea* (L.) Holub.; *Thelycrania sericea* (L.) Dandy

A cultivated shrub sometimes planted on roadsides.

ARALIACEAE

HEDERA L.

Hedera helix L. Ivy

Climbing on trees, rocks and walls and on the ground in woods and on hedgebanks. Very common almost everywhere in the county.

UMBELLIFERAE

HYDROCOTYLE L.

Hydrocotyle vulgaris L. Marsh Pennywort

In bogs and marshes particularly on acid soils. Common on Exmoor and the other hills of the west and south and on the peat moors (8). Rather rare elsewhere. It occurs in a few places on the Levels where the underlying peat is exposed in districts 3, 5 and 9, in the marsh at Berrow (9), at some marshy spots on the Mendip sandstone (9 & 10) and on the Wiltshire border (7 & 10). Recorded from all districts but absent from much of the county including the limestone areas.

SANICULA L.

Sanicula europaea L. Sanicle

In woods. Common over most of the county but absent from much of the Levels where there is no suitable habitat.

ASTRANTIA L.

Astrantia major L.

An occasional garden escape.

ERYNGIUM L.

Eryngium maritimum L. Sea Holly

Sandy sea shores. Formerly common, abundant in places. The number of plants started to decline about the turn of the century so that by 1912

UMBELLIFERAE

White wrote that little survived except in the more remote spots. Now very rare and only between Stolford (2) and Uphill (9). Of four reports during the last ten years two speak of a single plant only.

†**Eryngium campestre** L. Field Eryngo

Formerly in limestone grassland on Worlebury Hill, Weston-super-Mare (9), where it was first recorded in 1843. White saw it last in 1895 and it disappeared a few years later.

CHAEROPHYLLUM L.

Chaerophyllum temulentum L. Rough Chervil

C. temulum L.

Hedgebanks, roadsides and wood borders. Very common throughout the county.

ANTHRISCUS Pers.

Anthriscus caucalis Bieb. Bur Chervil

A. neglecta Boiss. & Reut.; *A. vulgaris* Pers.; *Chaerophyllum anthriscus* (L.) Crantz

In waste places, particularly on sandy ground, near the sea. In a few spots near the coast between Minehead (2) and Sand Point (9), sparse in most but quite plentiful between Burnham (8) and Brean (9). A rare casual further inland.

Anthriscus sylvestris (L.) Hoffm. Cow Parsley

Chaerophyllum sylvestre L.

On hedgebanks and roadsides. Abundant throughout the county.

†**Anthriscus cerefolium** (L.) Hoffm. Garden Chervil

A garden escape formerly quite widespread but not recorded since 1922.

SCANDIX L.

Scandix pecten-veneris L. Shepherd's Needle

A weed in cornfields and other cultivated land. Formerly common and generally distributed but now rare due to the use of weedkillers. Recently recorded only near Hinkley Point (2), at East Lambrook (4), at several spots on Sedgemoor (5), near Butleigh (8) and a few places around Bath (10).

MYRRHIS Mill.

Myrrhis odorata (L.) Scop. Sweet Cicely

Only an uncommon garden escape in Somerset.

TORILIS Adans.

Torilis japonica (Houtt.) DC. Upright Hedge-parsley

T. anthriscus (L.) C. C. Gmel.; *Caucalis anthriscus* (L.) Huds.

Roadsides, hedgebanks and wood borders. Very common everywhere except in the extreme west on Exmoor.

Torilis arvensis (Huds.) Link Spreading Bur-parsley

T. infesta sensu White; *Caucalis arvensis* Huds.

A weed of cornfields and arable land never as common in Somerset as in some areas but formerly widespread and recorded in all districts except 1 and 7. It has now become very rare and the only recent records are at Curry Rivel (4), 1978 (RGR), Greinton (5), 1972 (CAH), North Barrow (5), 1975 (CAH), Charlton Mackrell (5), 1978 (R & IR), and Meare (8), 1973 (CAH).

Torilis nodosa (L.) Gaertn. Knotted Bur-parsley

Caucalis nodosa (L.) Scop.

On dry banks, roadsides and in waste places. Quite frequent near the sea from Watchet (2) to the Avon estuary (10) and on the Lias in districts 3, 5 and 8. Rare elsewhere and only recorded recently at Chiselborough (4), Wedmore (8) and a few scattered localities in district 10.

CAUCALIS L.

†**Caucalis platycarpos** L. Small Bur-parsley

C. lappula Grande; *C. daucoides* L.

Once a rare cornfield weed on calcareous soils but it has not been recorded as such this century. There are a number of later records as a casual in gardens and waste places, the last being at Winscombe (9) in 1940.

UMBELLIFERAE

TURGENIA Hoffm.

†**Turgenia latifolia** (L.) Hoffm. Greater Bur-parsley

Caucalis latifolia L.

Another former cornfield weed recorded "on the hills at Bathe" in Gerard (1633) and at many places, particularly in district 5, up to the middle of the nineteenth century. It was last seen in a cornfield at Keynsham (10) in 1870 (T. B. Flower) and as a casual on waste ground at Burnham (8) in 1927 (WDM).

CORIANDRUM L.

Coriandrum sativum L. Coriander

An uncommon casual.

SMYRNIUM L.

Smyrnium olusatrum L. Alexanders

Formerly cultivated as a potherb but now well established on rocks, roadsides and waste ground particularly near the sea. Common near the coast in district 2 and frequent in the lowlands of districts 3, 4, 5, 8 and 9. There are very few records more than six or seven miles from the sea. Not recorded in districts 1, 6 or 7. The first British record was on Steep Holm (Turner, 1562) and its abundance there has been remarked on by many visiting botanists since. It has been checked from time to time by farming or other human activities but it now covers almost the whole of the plateau on the top of the island as well as being plentiful on the slopes and cliffs.

CONIUM L.

Conium maculatum L. Hemlock

On hedgebanks, by streams, on wood borders and in waste places. Very common except on the higher ground of Exmoor and the Quantock, Blackdown and Mendip Hills. Not recorded recently in district 1.

BUPLEURUM L.

†**Bupleurum rotundifolium** L. Thorow-wax

Formerly a cornfield weed on calcareous soil, particularly the Lias of district 5. It was recorded near Somerton (5) by Turner (1568) and persisted at Charlton Mackrell (where it was collected by Gapper in 1822) until at least 1948. A few more recent reports from other places are suspect; where specimens have been seen they have all proved to be *B. subovatum*.

Bupleurum tenuissimum L. Slender Hare's-ear

In salt marshes by the coast and tidal estuaries. Rather rare and local. In several places between Lilstock and Steart (2) and by the estuaries of the Parrett (3), Brue (8), Axe (9), Yeo (9) and Avon (10). In September 1977 it was found in some quantity in poor dry grassland on the southern slope (calcareous clay) of the Polden Hills in Stawell parish (5) (RSC) about five miles from the nearest salt water, a very unusual locality.

Bupleurum subovatum Link ex Sprengel False Thorow-wax

B. lancifolium auct.; *B. protractum* sensu White

A casual from the Mediterranean region first recorded in the county at Portishead (9) in 1906. It has become more frequent in recent years being found in gardens and on rubbish tips derived from imported bird-seed and has now been recorded in all districts except 1, 2, 5 and 6. There has been much confusion over the correct name for this species. The true *B. lancifolium* Hornem., a closely-allied species with smaller fruit, does not seem to have been identified in Britain yet but this is the name most often used at present.

TRINIA Hoffm.

Trinia glauca (L.) Dumort. Honewort

T. vulgaris DC.; *Apinella glauca* (L.) Caruel

In dry limestone turf. Rare but locally plentiful on Mendip from Axbridge to Uphill, where it was first discovered by Dillenius about 1726. Also on Worle Hill and Sand Point. All these localities are in district 9. It has never yet been found on Brean Down (9) and has not been recorded on the Somerset side of the Avon Gorge (10) since 1849. In 1868 Dr H. F. Parsons found a single minute plant which he thought might be this on the chalk of Long Knoll (7) (Wilts); the specimen in the Taunton herbarium is inadequate for identification.

UMBELLIFERAE

Cuminum L.

Cuminum cyminum L.

An uncommon casual first recorded in the county on a rubbish tip at Brislington (10), 1978 (ALG).

Apium L.

Apium graveolens L. Wild Celery

In marsh ditches, often in brackish water, especially near the sea. Locally common along the coast and up the tidal river estuaries between Minehead (2) and the Avon (10). Rare further inland in ditches of the Levels of districts 3, 4, 5, 8 and 9. Isolated occurrences elsewhere are probably derived from cultivated varieties.

Apium nodiflorum (L.) Lag. Fool's Water-cress

In ditches, by streams and in marshes. Very common in all districts.

Apium inundatum (L.) Reichb.f. Lesser Marshwort

In shallow pools and ditches. Very rare and uncertain in its appearances. Only seen recently in two pools on Mendip near Priddy (9), 1965 (CAH & C. R. Lancaster) and 1968 (R & IR). Known on the peat moors (8) to Sole (1782) it was not recorded there again until found by White a century later after which it was sighted from time to time until 1941. The only other records were near Weston Zoyland (5) (Marshall, 1914), Beacon Hill (10), 1926 (IMR), and Rudge (10) (H. F. Parsons in Murray, 1896) but it did not persist in any of these places.

Petroselinum Hill

Petroselinum crispum (Mill.) A. W. Hill Garden Parsley

P. sativum Hoffm.; *Carum petroselinum* (L.) Benth.

An escape from cultivation which can become established on rocks and old walls. In an old quarry at Bleadon (9) it has been known since 1889 and it is plentiful on rocks and walls in the Bristol Dock area between the old (Floating Harbour) and new (New Cut) courses of the Avon (10) (V.C.34). Recorded recently from all districts except 1, 2, 4 and 6, but in most places only as an obvious escape.

Petroselinum segetum (L.) Koch — Corn Parsley

Carum segetum (L.) Benth. ex Hook. f.

On grassy banks, roadsides and in arable land on calcareous soils. Now rare as a cornfield weed but quite frequent and perhaps increasing in its other habitats, particularly on the Lias of districts 3, 4, 5 and 8, the Bath Oolite (10), on river banks and near the sea. Very rare in the west of the county and not recorded in districts 1, 6 or 7.

SISON L.

Sison amomum L. — Stone Parsley

In hedges and on roadsides and railway banks. Common over much of the county especially in the lowlands. Absent from higher ground and not recorded recently from district 1.

CICUTA L.

† **Cicuta virosa** L. — Cowbane

In shallow water, marshes and ditches. Recorded by Sole (1791) as abundant on Burtle Moor (8) it became very rare and was last seen there in 1888 by H. S. Thompson. There are other old records from Burnham (8), Easton (9) and Bathampton (10) but it has long since disappeared from all these places.

AMMI L.

Ammi majus L. — Bullwort

An uncommon casual, first recorded in Somerset on a tip at Bedminster (10), 1977 (ALG).

Ammi visnaga (L.) Lam. — Toothpick Plant

Bird-seed alien.

CARUM L.

Carum carvi L. — Caraway

A rare casual.

UMBELLIFERAE

CONOPODIUM Koch

Conopodium majus (Gouan) Loret — Pignut

C. denudatum Koch; *Carum flexuosum* Fries

In woods and grassland. Very common over much of the county but rare in the Levels.

PIMPINELLA L.

Pimpinella saxifraga L. — Burnet Saxifrage

In dry grassland particularly on calcareous soils. Very common except on Exmoor where it is rare. Recorded in all districts.

Pimpinella major (L.) Huds. — Greater Burnet Saxifrage

P. magna L.

On hedgebanks. Very rare. It was found at North Cadbury (5) by A. L. Still in 1903 and still survives there. The only other occurrence in Somerset is referred to by White (1912) as an escape from C. E. Broome's garden at Batheaston (10); the plant was reported from the neighbourhood as lately as 1926 (WDM) but now seems to have gone.

AEGOPODIUM L.

Aegopodium podagraria L. — Gout-weed

On hedgebanks, by roadsides and a pernicious weed in gardens. Very common throughout the county but seldom far from houses.

SIUM L.

Sium latifolium L. — Greater Water Parsnip

In marsh ditches and by rivers. Formerly plentiful on the Levels but now rather rare in districts 4, 5 and 8 only. Not seen recently in districts 3 or 9.

BERULA Koch

Berula erecta (Huds.) Coville — Lesser Water Parsnip

Sium erectum Huds.

In ditches, streams, ponds and canals. Common throughout the Levels and in the Kennet & Avon canal (10). Rare elsewhere though recorded in all districts except 1 and not recently in 7.

CRITHMUM L.

Crithmum maritimum L. Rock Samphire

On rocks, sea walls and in sandy places by the sea. Formerly common but apparently decreasing. Only seen recently in district 2 near Porlock and at Hinkley Point. More frequent in district 9 between Berrow and Portishead. Still on Steep Holm where it was recorded by Lightfoot in 1773.

OENANTHE L

Oenanthe fistulosa L. Tubular Water Dropwort

In marsh ditches and by ponds. Common throughout the Levels in districts 3, 4, 5, 8 and 9. Very rare elsewhere. Not recorded in districts 1 or 6 and not seen recently in 2, 7 or 10.

Oenanthe pimpinelloides L. Corky-fruited Water Dropwort

In damp meadows and on roadside verges. Common in the south of the county in districts 3, 4, 5, 6, 7 and 8. Rather rare in the north and west and not recorded recently in district 1.

(**Oenanthe silaifolia** Bieb (*O. peucedanifolia* auct.). A record published by White (1886) was said by him in his later book (1912) to have been an error.)

Oenanthe lachenalii C. C. Gmel. Parsley Water Dropwort

In marshes, especially near the sea. Common near the coast between Stolford (2) and Clevedon (9) and on the Levels for a few miles inland. Very rare elsewhere and usually in much wetter places than *O. pimpinelloides* with which it is sometimes confused. There are several recent records round Bath (10) but otherwise it has only been seen well away from the coast at Haselbury (4), 1966 (EBH & CAH), and Hinton Blewitt (10), 1973 (JA). Not recorded in districts 1 or 6 and not recently in 7.

Oenanthe crocata L. Hemlock Water Dropwort

By ditches, streams and in wet woodland. Very common everywhere except on the higher parts of Mendip.

UMBELLIFERAE

Oenanthe aquatica (L.) Poir. Fine-leaved Water Dropwort

O. phellandrium Lam.

In marsh ditches. Common in the Levels of central Somerset in districts 3, 4, 5 and 8. Rather rare in those of district 9.

Oenanthe fluviatilis (Bab.) Colem. River Water Dropwort

In rivers, streams and canals. Very rare and only seen recently in the Rivers Isle and Parrett near Drayton (4), Cary and its tributaries near Somerton (5) and in the Kennet & Avon canal at Bath (10).

Aethusa L.

Aethusa cynapium L. Fool's Parsley

A very common weed of arable fields and gardens.

Foeniculum Mill.

Foeniculum vulgare Mill. Fennel

F. officinale All.

On rocks and in waste places. Frequent and perhaps native near the coast and in the Cheddar valley (9). In other places further inland usually near houses and a relic of cultivation. Not recorded in districts 1, 6 or 7.

Anethum L.

Anethum graveolens L. Dill

A rare or overlooked casual first recorded in the county at Brislington (10), 1917 (CIS), and seen there again in 1978 (ALG) and also at Stawell (5), 1978 (CES).

Silaum Mill.

Silaum silaus (L.) Schinz & Thell. Pepper Saxifrage

Silaus flavescens Bernh.; *S. pratensis* Bess.

In grassland, particularly on clay. Common in the eastern half of the county but rare in the west. Not recorded in district 1.

ANGELICA L.

Angelica sylvestris L. Wild Angelica

By ditches, on streamsides and in damp woods. Very common throughout the county.

PEUCEDANUM L.

Peucedanum palustre (L.) Moench Milk Parsley

In fens. Very rare and only on the peat moor (8). Plentiful once according to Sole (1791) but now scarce and restricted to a small area.

PASTINACA L.

Pastinaca sativa L. Wild Parsnip

Peucedanum sativum (L.) Benth. ex Hook. f.

Grassland and waste places. Very common on calcareous soils in the eastern half of the county. Rare in the west except on the Lias of district 2. Not recorded in districts 1 or 6.

HERACLEUM L.

Heracleum sphondylium L. Hogweed

On roadsides and in grassy and waste places. Very common throughout the county. The var. **angustifolium** Huds. is quite frequent, particularly in the eastern part of district 8 and in district 10.

Heracleum mantegazzianum Somm. & Levier Giant Hogweed

H. giganteum auct.

An ornamental plant from south-west Asia which has become established in a number of places scattered about the county but shows no sign of spreading as it has done in some other parts of the country.

DAUCUS L.

Daucus carota L. Wild Carrot

In grassland and on roadsides. Very common in most of the county especially on calcareous soils but rare on Exmoor and the Quantocks. Not recorded recently in district 1.

ssp. **gummifer** Hook. f. (*D. gummifer* Lam.) was recorded from Shurton Bars (2) by J. C. Collins and a specimen collected there in 1915 by H.

Slater is in the herbarium at Taunton. White did not accept early records from Brean Down (9) but Marshall found it there in 1914 and was supported in its occurrence there by the Rev E. Ellman in 1919. It has not been positively identified in either of these localities recently.

CUCURBITACEAE

Bryonia L.

Bryonia cretica L. ssp. **dioica** (Jacq.) Tutin — White Bryony

B. dioica Jacq.

Hedges and wood borders. Although recorded in all districts except 1 and 7 its distribution is very uneven. From Lilstock (2) along the eastern side of Quantock to Taunton (3), Langport (5) and Yeovil (4) it is quite common, and again in a broad belt in districts 9 and 10 from Loxton and Cheddar north to Clevedon and Portishead and along the county border near Bristol and Bath to Frome. Outside these areas it is rare and only in a few isolated localities. In Somerset this species is at its western limit as an English native but the factors which determine its peculiar distribution are far from obvious, particularly its absence from areas where the soil and other conditions are apparently the same as in adjacent areas where it is frequent.

Citrullus Schrader

Citrullus lanatus (Thunb.) Mansfeld — Water Melon

Sometimes found on rubbish tips, as at Brislington (10), 1978 (ALG).

Sicyos L.

Sicyos angulatus L. — One-seeded Bur Cucumber

Occurred as a garden weed at Wrington (9) in 1964 (W. G. Sims).

ARISTOLOCHIACEAE

Asarum L.

†**Asarum europaeum** L. — Asarabacca

Said to have been found in Somerset in the seventeenth century but if this was correct it has long become extinct.

EUPHORBIACEAE

MERCURIALIS L.

Mercurialis perennis L. Dog's Mercury

In woods and on shady hedgebanks. Very common throughout the county except in those parts of Exmoor and the Levels where no suitable habitat exists.

Mercurialis annua L. Annual Mercury

A weed in gardens and waste places. Very common since Babington's day about Bath (10) it has since spread to most other parts of the county though still rare in the south and not recorded from districts 1 or 6. First recorded on Steep Holm in 1952 it has now become abundant there.

EUPHORBIA L.

†**Euphorbia peplis** L. Purple Spurge

On sandy shores. Recorded by Gapper and Collins in the *New Botanist's Guide* and *Supplement* (1835-37) it has not been seen by any later botanist but it has declined to extinction elsewhere in Britain and there is no reason to doubt its former occurrence in Somerset.

Euphorbia lathyris L. Caper Spurge

A weed in gardens and waste places. An escape or survival from cultivation it has been recorded as a casual in all districts but its long persistence in two localities has suggested it might be a native. Recorded by Banks and Lightfoot on Steep Holm in 1773 it still survives there. Babington (1839) thought it might be indigenous in Warleigh Wood near Bath (10) and it is still there though varying much in quantity from year to year; only one small patch could be found in 1979 (DG).

†**Euphorbia villosa** Waldst. & Kit. ex Willd. Hairy Spurge

E. pilosa auct.

Formerly in a wood and a lane hedge at Bath (10) where it was first recorded by Lobelius (1576). White (1912) described it as scattered over about two acres of coppiced woodland in 1884 but as diminishing in later years with the abandonment of coppicing till by 1909 it was much reduced by other herbage. It maintained a struggling existence till by 1941 T. H. Green said only one plant remained; it has not been recorded since.

EUPHORBIACEAE

†**Euphorbia hyberna** L. Irish Spurge

Formerly in woods on the Somerset side of the Badgworthy Water (2) but it has not been recorded there since 1929.

Euphorbia platyphyllos L. Broad-leaved Spurge

Cornfield weed. Always rare and local in most of the county it is decreasing and is now almost confined to the Lias of central Somerset in districts 5 and 8 where it is still quite frequent. It has not been recorded recently outside this area except at Stockland Bristol (2), 1971 (HWB), and at three spots between Bath and Frome (10). It was formerly recorded in districts 3, 4 and 9 also. The first British record was at Keynsham (Ray, 1670) but it had not been seen in that area since 1920 until it was found on a tip at Brislington (10), 1979 (ALG & OMS).

Euphorbia serrulata Thuill. Upright Spurge

E. stricta L.

This native of the Wye valley was first seen on a field border in Bathampton (10) in 1947 (Misses A. L. & J. D. Miller) and still survives there. It presumably originated from a near-by garden. It had previously been recorded as a casual at Twerton (10) in 1902. Old records in Babington (1834) probably referred to *E. platyphyllos.*

Euphorbia helioscopia L. Sun Spurge

A weed of arable land. Very common throughout the county except on the moors of the west where it is rare.

Euphorbia peplus L. Petty Spurge

A weed of gardens and cultivated land. Very common and recorded in all districts.

Euphorbia exigua L. Dwarf Spurge

In cornfields and in other cultivated ground. Apparently decreasing in Somerset. Described by Murray (1896) as very common, recent records are widely but rather thinly scattered about the county mainly on calcareous soils. Rather rare in the west and not recorded recently in district 1.

Euphorbia paralias L. Sea Spurge

Sandy shores. Local and decreasing on the coast from Stert Island (2) to Brean (9). Formerly at Minehead (2) but not recorded there since 1953, and at Weston-super-Mare (9) where it was already scarce in 1912 according to White but has since disappeared.

Euphorbia esula L. sensu lato Leafy Spurge

An occasional casual of railway banks and waste places. Since 1928, when it was recorded at Ashton Gate (10) (CIS), it has been reported in several other places in the north of the county in districts 9 and 10. The aggregate is believed to consist of two subspecies, ssp. **esula** and ssp. **tommasiniana** (Bertol.) Nyman (*E. uralensis* Fisch. ex Link; *E. virgata* Waldst. & Kit.), and the hybrid between them. The usual British plant is this hybrid (*E.* × *pseudovirgata* (Schur) Soó). Ssp. *esula* has occurred in Britain, mainly in the north and in the nineteenth century, and Flower's old record near Farleigh Castle (10) in Babington (1839) may have been of this.

Euphorbia amygdaloides L. Wood Spurge

In woods and occasionally in hedges. Common in districts 3 and 4 and again in the north of the county in districts 8, 9 and 10. Rare elsewhere particularly on Exmoor and in the Levels where there is little woodland. Not recorded recently in district 1.

POLYGONACEAE

POLYGONUM L.

Polygonum aviculare L. Knotgrass

Cultivated ground, roadsides and waste places. Very common throughout the county.

Polygonum arenastrum Bor.

P. aequale Lindm.; *P. calcatum* Lindm.

The distribution of this segregate of *P. aviculare* sensu lato has not been completely worked out but it seems to be almost equally as ubiquitous as *P. aviculare* sensu stricto.

POLYGONACEAE

Polygonum rurivagum Jord. ex Bor.

As yet there have only been a few rather doubtful records of this segregate in Somerset and its occurrence in the county has not been definitely established.

Polygonum oxyspermum Mey. & Bunge ex Ledeb. ssp. **raii** (Bab.) D. A. Webb & Chater — Ray's Knotgrass

P. raii Bab.; *P. robertii* auct.

Sandy shores. Very rare and only seen recently on Stert Island (2), 1974 (RSC) and at Berrow (9). Formerly recorded also at Dunster (2), Steart (2) and Burnham (8).

†**Polygonum maritimum** L. — Sea Knotgrass

A single plant of this very rare and decreasing species was found on the shore between Burnham (8) and Brean (9) in July 1882 by H. S. Thompson; the specimen is now at Kew. St Brody (1856) recorded it as "rarely on the beach at Weston-super-Mare". It has been suggested that this might have been a mistake for *P. oxyspermum* but neither species has been seen there since and there is no reason to doubt St Brody's identification.

Polygonum bistorta L. — Common Bistort

In damp meadows and on roadsides and occasionally in woods. Frequent in the west and south and at the eastern end of Mendip. Rare elsewhere and sometimes only a garden escape. Decreasing in the north where it has gone from many of its old localities. Absent from a wide tract in the centre of the county including the whole of district 5.

Polygonum amplexicaule D. Don — Red Bistort

Garden escape.

Polygonum amphibium L. — Amphibious Bistort

In and by ditches, pools, streams and rivers, the terrestrial form sometimes a weed in arable land. Very common throughout the Levels. Frequent on the lower ground elsewhere except in the west where it is rare. Not recorded in district 1.

Polygonum persicaria L. Redshank

A weed of arable land and waste places. Very common in all districts.

Polygonum lapathifolium L. Pale Persicaria

A weed of arable land and waste places. Common in the eastern half of the county, less so in the west and not recorded in district 1. **Polygonum nodosum** Pers. (*P. maculatum* (Gray) Dyer ex Bab.) is a fairly frequent variety usually found on damper ground than the type.

Polygonum hydropiper L. Water-pepper

By ditches and streams and wet woodland tracks. Very common throughout the county.

Polygonum mite Schrank Tasteless Water-pepper

In damp places. Very rare or overlooked. It has only been recorded recently at Hambridge (4), 1970 (JGK), and at a few spots on the peat moors (8). Formerly also at Weston Zoyland (5) and Compton Dando (10).

Polygonum minus Huds. Small Water-pepper

In wet places. Very rare and only recorded recently at Meare Heath (8), 1960 (Dr Munro Smith). Seen on the peat moors by both Sole and Gapper it was not found again till 1895 (JWW) since when it has been recorded there from time to time. The only other Somerset records were at Stogumber (2), Gasper (7), Priddy (9) and Nailsea (9) but it was never found again in any of these places.

Polygonum polystachyum Wall. ex Meisn. Himalayan Knotweed

A shrubby species from the Himalayas first recorded in the county as an escape near Clevedon (9) in 1943 (C & NS). Since then it has become established by roadsides and in waste places in a number of localities scattered about the county in all districts except 3.

Polygonum campanulatum Hook. f. Lesser Knotweed

Garden escape, first recorded in Somerset at Spaxton (3), 1974 (HWB).

Polygonum nepalense Meisn.

A casual once found on a rubbish tip at Yeovil (4), 1963 (VIR det. CCT).

POLYGONACEAE

Polygonum patulum Bieb.

P. bellardii auct.

First found in the county at Ashton Gate (10), 1930 (CIS), this casual also appeared at Yeovil (4), 1963 (VIR det. CCT).

BILDERDYKIA Dum.

Bilderdykia convolvulus (L.) Dum. Black Bindweed

Polygonum convolvulus L.; *Fallopia convolvulus* (L.) A. Löve

In cornfields and other cultivated ground and sometimes in waste places. Very common throughout the county.

†**Bilderdykia dumetorum** (L.) Dum. Copse Bindweed

Polygonum dumetorum L.; *Fallopia dumetorum* (L.) Holub

In hedges and thickets. Formerly by the railway near Keynsham (10) where it was said by Flower to have been discovered by D. Don. Specimens collected by Babington in 1836 are in the herbaria at Kew, Bristol and Warwick, but it had gone by the end of the century.

Bilderdykia aubertii (L. Henry) Moldenke Russian Vine

Polygonum aubertii L. Henry; *P. baldshuanicum* auct.; *Fallopia aubertii* (L. Henry) Holub

A vigorous introduced climber which has become established in places but seldom far from gardens.

REYNOUTRIA Houtt.

Reynoutria japonica Houtt. Japanese Knotweed

Polygonum cuspidatum Sieb. & Zucc.

An introduced species which can become a pest in gardens. First recorded in Somerset as a weed of waste places at Portishead Dock (9), 1951 (C & NS), it is now frequent and widespread in all parts of the county. A dwarf variety, var. **compacta** (Hook. f.) Buchheim, was recorded as an escape at Arno's Vale, Bristol (10) in 1945 (IWE) but has not been noticed anywhere else.

REYNOUTRIA

Reynoutria sachalinensis (F. Schmidt) Nakai — Giant Knotweed

Polygonum sachalinense F. Schmidt

Much less common and aggressive than the last species it was first recorded in Somerset on a waste tip at Frome (10) in the 1950s (PFH). Since then it has been recorded in thirteen other localities, mostly in districts 8, 9 and 10, with single spots in districts 2, 6 and 7.

FAGOPYRUM Mill.

Fagopyrum esculentum Moench — Buckwheat

F. sagittatum Gilib.

Sown as food for pheasants and occasionally found as a casual.

RUMEX L.

Rumex acetosella L. sensu lato — Sheep's Sorrel

In dry and rocky grassland, heaths and waste places, particularly on acid soils. Common over much of the county, especially in the west, and recorded from all districts but sparse in the lowlands and on the eastern Oolite. The distributions of the segregates **R. angiocarpus** Murb. and **R. tenuifolius** (Wallr.) Löve has not been worked out but both are believed to occur in Somerset.

Rumex acetosa L. — Common Sorrel

In meadows and pastures. Very common throughout the county.

Rumex hydrolapathum Huds. — Water Dock

By rivers, canals, moor ditches and ponds. Common throughout the Levels in districts 3, 4, 5 and 8, less so in 9. Frequent by the Kennet & Avon canal and River Avon in district 10. Outside these areas it is rare and often by ornamental water where it may have been originally planted. Not recorded in district 1 nor recently in 7.

× **obtusifolius** (*R.* × *weberi* Fisch.-Benz.) has been recorded in the county, most recently on Walton Heath (8), 1947 (C & NS conf. Dr K. H. Rechinger).

Rumex crispus L. — Curled Dock

Cultivated ground, grassland and waste places. Very common throughout the county.

× **obtusifolius** (*R.* × *pratensis* Mert. & Koch; *R.* × *acutus* auct.) appears to be not uncommon by roadsides and in other disturbed ground and has been recorded in all districts except 1 and 6.

Rumex obtusifolius L. Broad-leaved Dock

By roadsides, in fields and waste ground. Very common everywhere.

× **sanguineus** (*R.* × *dufftii* Hausskn.) has not been identified recently but is probably overlooked. The last confirmed record was from Hembury Hill (8 or 9) near Wookey, 1944 (JPMB conf. JEL).

Rumex pulcher L. Fiddle Dock

In dry thin turf particularly on limestone hillsides and by roadsides. Frequent on the Lias of districts 2 and 5, the Carboniferous Limestone of 8 and 9 and the Oolite of 10. Rare elsewhere. Not recorded in districts 1 and 6 nor recently in 4 and 7.

Rumex sanguineus L. Wood Dock

In woods, lanes and waste places. Very common throughout the county.

Rumex conglomeratus Murr. Clustered Dock

By ditches, in marshes and wet fields and on roadsides. Very common in all districts.

× **hydrolapathum** (*R.* × *digeneus* G. Beck) was found on a bank of a rhine on Walton Heath (8) in 1942 (C & NS conf. Dr K. H. Rechinger.)

× **crispus** (*R.* × *schulzei* Hausskn.) has been recorded at Chedzoy (5), 1907 (C. E. Salmon & ESM), and Taunton (3), 1945 (WW), but is probably more frequent than these meagre records imply.

× **obtusifolius** (*R.* × *abortivus* Ruhmer) has been recorded only once, at Burton Pynsent (3), 1930 (WW).

× **pulcher** (*R.* × *muretii* Hausskn.) has been recorded in V.C.6.

Rumex palustris Sm. Marsh Dock

R. limosus auct.

In marshes and by ditches and streams. Rare and local. Still to be found on West Sedge Moor (3), Sedge Moor (5) and, more plentifully, in abandoned peat cuttings on the peat moors (8). Formerly more widespread but it has not been seen on the moors near Nailsea (9) since 1941. Some of the old records were probably of casual occurrences as were two recent ones at Hawkridge (2) and Weston-super-Mare (9).

Rumex maritimus L. Golden Dock

In marshes. Very rare and only seen recently in old peat cuttings on the moors near Shapwick and Westhay (8). Formerly more widespread on the Levels in districts 3, 4, 5 and 9 also.

Rumex brownii Campd.
Rumex obovatus Danser

Wood adventives recorded on a tip at Glastonbury (8), 1970 (CAH, CES & JGK det. JEL).

URTICACEAE

PARIETARIA L.

Parietaria judaica L. Pellitory-of-the-wall

P. diffusa Mert. & Koch; *P. officinalis* auct.; *P. ramiflora* auct.

On walls, dry banks and sometimes on rocks. Common in all parts of the county except on Exmoor. Plentiful on the cliffs of Steep Holm where it was first recorded by T. Clark in 1831.

SOLEIROLIA Gaud.-Beaup.

Soleirolia soleirolii (Req.) Dandy Mind-your-own-business
Helxine soleirolii Req.

A frequent garden escape but seldom far from houses.

URTICA L.

Urtica urens L. Small Nettle

A weed of arable land, gardens and waste places. Common over much of the county except for Exmoor. Abundant on Steep Holm. Not yet recorded for district 1.

Urtica dioica L. Common Nettle

Woods, hedgebanks, about farms and in waste places. Abundant in all parts of the county.

URTICACEAE

†Urtica pilulifera L. Roman Nettle

A very rare casual only once recorded in the county, at Kewstoke (9) by St Brody (1856).

CANNABACEAE

HUMULUS L.

Humulus lupulus L. Hop

In hedges and wood borders. Common throughout the county.

CANNABIS L.

Cannabis sativa L. Hemp

A casual, usually derived from bird seed, increasingly found on rubbish tips.

ULMACEAE

ULMUS L.

Ulmus glabra Huds. Wych Elm

U. montana With.

In woods and hedges. Very common and an important constituent of woodland on the limestone in the north of the county. Less common and often planted elsewhere but widely distributed in all districts. In the north of the county where Dutch elm disease is so prevalent some trees have succumbed to it but it is not so susceptible as the next species.

Ulmus procera Salisb. English Elm

U. campestris auct.

Hedgerows, fields and copses. Once abundant throughout the county except on Exmoor but Somerset was one of the first areas to be struck by Dutch elm disease, which may have arrived from North America via the port of Bristol about 1969, and the majority of fully-grown trees have died. There are still saplings in the hedges and it remains to be seen whether the species will survive the epidemic.

Ulmus minor Mill. Small-leaved Elm

U. carpinifolia Gled.; *U. stricta* (Ait.) Lindl.

Not native in Somerset but occasionally planted.

Ulmus × hollandica Mill. Dutch Elm

A frequently planted hybrid tree believed to be derived from *U. glabra, U. minor* and *U. plotii* Druce, itself possibly a variant of *U. minor*.

MORACEAE

MORUS L.

Morus alba L. White Mulberry

A single young tree in a planted hedge of Hornbeam was found near West Harptree (10) in 1968 by the Rev H. D. Wiard and Mrs M. Webber; no doubt a nurseryman's error.

FICUS L.

Ficus carica L. Fig

An occasional casual. A bush, already some years old, was first noticed on Steep Holm in 1964 (VG) and still survives there.

JUGLANDACEAE

JUGLANS L.

Juglans regia L. Walnut

Often self-sown in hedges, on railway banks and in waste places.

MYRICACEAE

MYRICA L.

Myrica gale L. Bog Myrtle

In peaty bogs. Rare and local. Churchstanton (6) (V.C.3) and Widcombe (6) Moors, Chard Common (6) and in several places on the peat moors (8). Formerly recorded near East Anstey (1), Staple Fitzpaine (4) and Easton (9) but not seen in any of these places for many years. According to Sole (1791) it once grew on King's Sedge Moor (5) too.

PLATANACEAE

Platanus L.

Platanus hybrida Brot. London Plane

P. acerifolia (Ait.) Willd.

Often planted. There are no records of self-sown trees in Somerset. The origin is uncertain; it could be a cultivated variety of *P. orientalis* L. or its hybrid with some other species.

BETULACEAE

Betula L.

Betula pendula Roth Silver Birch

B. verrucosa Ehrh.; *B. alba* auct.

In woods, particularly on hilly acid soils. Commonly planted elsewhere. Recorded in all districts but doubtfully native in the lowlands.

Betula pubescens Ehrh. Downy Birch

B. tomentosa sensu White

On damp heaths and wet peaty acid soils. Seldom planted so more confined to the moors of the west and south, the peat moors, the clays and Greensand of the eastern borders and the Gordano valley than *B. pendula*.

Puzzling intermediates between the two species of Birch occur frequently. These do not seem to be hybrids, which are difficult to produce artificially and are then completely sterile, but may be examples of the variability of the species, particularly *B. pubescens* which may itself be originally of hybrid origin.

Alnus Mill.

Alnus glutinosa (L.) Gaertn. Alder

By streams and rivers, in marshes and damp woods. Common throughout the county.

Alnus incana (L.) Moench Grey Alder

Has been planted near Westhay (8) and in the Gordano valley (9) but it is still very uncommon in Somerset.

CORYLACEAE

CARPINUS L.

Carpinus betulus L. Hornbeam

Widely planted in woods and hedges.

CORYLUS L.

Corylus avellana L. Hazel

Woods and hedges. Very common in all districts.

FAGACEAE

FAGUS L.

Fagus sylvatica L. Beech

Perhaps native in woods in the north of the county but widely planted in woods and hedges.

NOTHOFAGUS Blume

Nothofagus obliqua (Mirbel) Blume Roblé Beech

Planted at Kingswood Warren (8) in 1936/37. This and **Nothofagus procera** (Poeppig & Endl) Örsted, both from Chile, are now being widely planted in Britain but this does not seem to be happening yet in Somerset. They have been suggested as replacements for Elms killed by Dutch elm disease.

CASTANEA Mill.

Castanea sativa Mill. Spanish Chestnut

In woods and plantations. Originally planted but sometimes becoming naturalized especially on acid soils. Recorded from all districts but absent from the Levels and much of the limestone.

QUERCUS L.

Quercus cerris L. Turkey Oak

Originally planted in woods and park land it was first recorded as regenerating near Tickenham (9) in 1918. Now widely spread and in all districts but still rare in the west. In some areas it has become quite a pest forming scrub on previously open grassy slopes.

FAGACEAE

Quercus ilex L. Evergreen Oak

Widely planted but often regenerating in woods on the limestone and near the sea.

Quercus robur L. Common Oak

In woods and hedges. Very common throughout the county.

Quercus petraea (Mattuschka) Liebl. Sessile Oak

Q. sessiliflora Salisb.

Woods and hedges. The common oak of woods on Exmoor and the Quantocks. Also frequent on the Greensand on the Wiltshire border and in Leigh Woods (10). Elsewhere there are only a few widely scattered records.

× **robur** (*Q.* × *rosacea* Bechst.). Intermediates between the two native species have been seen but they are not as frequent as in some other parts of the country. Whether these are hybrids is far from certain.

Quercus rubra L. Red Oak

Q. borealis Michx. f.; *Q. maxima* (Marsh.) Ashe

Planted for ornament and occasionally for forestry, as at Witham Park (10).

SALICACEAE

Populus L.

Populus alba L. White Poplar

Widely planted.

Populus canescens (Ait.) Sm. Grey Poplar

Widely planted but possibly native in damp woodland in the north of the county. Although long known in Britain this is now thought to be the hybrid between the native *P. tremula* and the introduced *P. alba* and may well have arisen spontaneously here. The tree known as **P. × hybrida** Bieb., an example of which was recorded at Corston (10) in 1938 (CIS), is probably a nothomorph of this cross or it could conceivably be a backcross of *P. canescens* with *P. tremula*.

Populus tremula L. Aspen

Damp woods and hedges. Rare in the west and, except as a planted tree, in the Levels, but frequent elsewhere.

Populus nigra L. Black Poplar

A special survey has located a number of trees, particularly in the Taunton (3) area, mostly by rivers and streams and well grown. These are probably survivors of the time when the species was quite often planted in the lowlands until it was superseded by the faster-growing *P. × canadensis*. A cultivar, the Lombardy Poplar (*P. italica* (Duroi) Moench) is still quite frequently planted.

Populus × canadensis Moench Italian Poplar

P. serotina Hartig

A hybrid between *P. nigra* and the American *P. deltoides* Marsh., which is very widely planted throughout the county particularly in the lowlands where, after the apple, it is one of the most frequent hosts of Mistletoe.

Populus gileadensis Rouleau Balsam Poplar

Occasionally planted as an ornamental tree.

SALIX L.

Salix pentandra L. Bay Willow

Only a planted tree in Somerset.

Salix alba L. White Willow

Very common throughout the Levels and frequent elsewhere. Usually planted but may be a native in some localities.

× **fragilis** (*S. × rubens* Schrank). Frequent throughout the Levels.

Salix fragilis L. Crack Willow

Very common, especially by water, throughout the county except in the west where it is rare. Often planted.

Salix triandra L. Almond Willow

By streams and in damp ground. Rather rare in the wild and only in the eastern half of the county. Extensively grown for basket work in the osier beds of central Somerset.

× **viminalis** (*S.* × *mollissima* Hoffm. ex Elwert.) was found between Berrow and Lympsham (9) in 1932 (CIS det. J. Fraser).

Salix purpurea L. Purple Willow

Formerly planted in osier beds on the Levels but now very seldom seen.

× **viminalis** (*S.* × *rubra* Huds.). A basket willow, the Green-leaved Osier, only once recorded in Somerset, near Minehead (2), 1906 (A. Ley).

Salix viminalis L. Osier

By ditches, streams, in withy beds and in damp waste places. Rare in the west but frequent elsewhere.

Salix capraea L. Goat Willow

In woods and hedges. Common in all districts.

× **atrocinerea** (*S.* × *reichardtii* A. Kerner). Intermediate forms are very common and are probably this or backcrosses with one of the parents.

Salix atrocinerea Brot. Sallow

S. cinerea L. ssp. *atrocinerea* (Brot.) Silva & Sobrinho

In hedges, woods and by streams. Very common throughout the county.

× **purpurea** × **viminalis** (*S.* × *forbyana* Sm.) A basket osier which has been recorded from near Berrow and Brent Knoll (9) but has not been seen recently.

× **viminalis** (*S.* × *smithiana* Willd.) appears to be fairly frequent in the Levels.

Salix aurita L. Eared Sallow

Moors, hedges and damp woods on acid soils. Frequent on Exmoor, the Brendon and Blackdown Hills, the Greensand of the Wiltshire border and on the Mendip sandstone. Not recently recorded elsewhere although the older records indicate it used to be more widespread.

× **viminalis** (*S.* × *fruticosa* Doell.). A plant identified as this by the Rev E. F. Linton was found in a hedge near Ilchester (5) by Murray in 1891.

× **capraea** (*S.* × *capreola* A. Kerner ex Anders.). Reported several times in the county. The most convincing was one from Simonsbath (1) in 1918 by Marshall.

× **cinerea** (*S.* × *multinervis* Doell.). Recorded from Simonsbath (1) in 1918 by Marshall and in Monk's Wood, Cranmore (10) in 1957 by N. Y. Sandwith.

Salix repens L. Creeping Willow

On wet heaths. Rare and reported recently only from near Exford and East Anstey (1), at Langford Heathfield (3), on the Blackdowns (3, 4 & 6), at Chard Common (6), on the peat moors (8), and near Blackslough and Witham (8 & 10). Formerly recorded at a few spots on Mendip but not seen there since 1953. Other places in the north from which it has disappeared are Nailsea (9), Brislington and Berkley (10).

Salix arenaria L.

S. repens L. ssp. *argentea* (Sm.) G. & A. Camus

In dune slacks. Very rare and only at Berrow (9).

ERICACEAE

RHODODENDRON L.

Rhododendron ponticum L. Rhododendron

Planted in woodland in all parts of the county, it has become fully naturalized in acid-soil areas, spreading on to the heathland of Exmoor, the Quantocks and Blackdowns. Less prolific in other areas but now recorded in all districts. There are no published records before 1932 but it clearly became established in the county long before that.

ANDROMEDA L.

Andromeda polifolia L. Bog Rosemary

Extremely rare and now only on the peat south of Wedmore (8). Abundant on the drier parts of the moor until at least 1862 it became very rare by the end of the century due to the increased and extensive peat cutting. From 1920, when it was last seen by Mrs C. I. & N. Y. Sandwith, it was not found again until 1970 when it was discovered by J. K. Hibberd in the course of an intensive survey of the peat moor vegetation. In 1918

Mrs. C. I. Sandwith found a few small patches under tussocky grass on Blackdown on Mendip (9). It has not been seen there since 1928 but may still survive.

ARBUTUS L.

Arbutus unedo L. Strawberry Tree

Originally introduced but regenerating in woods at Culbone (2), Goblin Combe (9) and in the Avon Gorge (10).

CALLUNA Salisb.

Calluna vulgaris (L.) Hull Ling

C. erica DC.

On heaths and commons on acid soils. On limestone only where locally acid conditions have developed through leaching. Abundant on the moors and hills of the west and south. Common on the peat moors (8), on the Greensand of the eastern border, and on sandstone and the limestone heaths of Mendip. Very rare elsewhere. Recorded recently only on limestone heath at Wrington Warren (9), on Old Red Sandstone at Leigh Woods (10) and on Pennant Sandstone at Clutton and Hunstrete (10). Never recorded in district 5.

ERICA L.

Erica tetralix L. Cross-leaved Heath

On wet heaths and commons. Plentiful on the moors and commons of the west and south. Elsewhere only on the peat moors (8) and the Mendip sandstone (9). Not seen recently at the eastern end of Mendip (10) where it was formerly recorded, and long since gone from Leigh Woods (10) where it was last seen by T. B. Flower. Not recorded in districts 5 or 7.

Erica cinerea L. Bell Heather

On heaths and commons in drier places than the last species. The distribution is very similar but rather more widespread. Scarce on the peat moors (8). Extends further east on Mendip (10) and occurs also on the Wiltshire border near Penselwood (7) and Witham (10), and on limestone heath on Wrington Warren (9). Not recorded in district 5.

VACCINIUM L.

Vaccinium vitis-idaea L. Cowberry

Moors. Extremely rare. Only in one small area at the northern end of the Quantock Hills (2). First found in 1917 by the Rev C. Q. Knowles, who collected specimens, it could not be re-located by anyone else. In 1958 A. D. & O. M. Hallam rediscovered it in a spot about a quarter of a mile from what was believed to be the original locality and it still survives there. Ancient records from Weston-super-Mare (9) (Rutter, 1829) and Leigh Woods (10) (*Botanist's Guide,* 1805, and Collins, 1837) are believed to have been errors.

Vaccinium myrtillus L. Whortleberry

On heaths and in woodland on acid soils. Abundant on Exmoor and the hills of the west and on the Blackdowns in the south. Common from Penselwood (7) north to Witham (10) and on the Mendip sandstone (9 & 10). Formerly on the sandstone near Portbury (9) and in Leigh Woods (10) but it had become rare by White's time and has not been recorded there since 1921. Not on the peat moor (8) nor in district 5.

Vaccinium oxycoccus L. Cranberry

Oxycoccos palustris Pers.

In wet bogs, usually with *Sphagnum* spp. Very rare and now only in the blanket bogs of Exmoor between the Devonshire border and Dunkery Hill (1 & 2). There are very old records for Selworthy (2) and Brendon Hill (3). More recently it was found near Churchstanton (6) (V.C.3) in 1915 (WW). Always rare on the peat moors (8) it was last seen there in 1919 (C & NS). Known on Mendip since 1860 a small patch on Blackdown (9) was last reported in 1957 (C & NS) and may still survive.

PYROLACEAE

PYROLA L.

†**Pyrola minor** L. Common Wintergreen

Formerly in Tetton Woods (3), in a wood near Failand (10) and in Leigh Woods (10) but there is no record of its having been seen in any of these places since the publication of White's *Bristol Flora* (1912).

PYROLACEAE

MONOTROPA L.

Monotropa hypopitys L. Yellow Bird's-nest

Hypopitys multiflora Scop.

In beech and fir woods. Very rare and only seen recently in woods at Warleigh and Bathford (10). There are old records for Dunster (2), Weston-super-Mare and Brockley (9) and more recent ones for Copley Wood (5) and Butleigh (8) (1943), Leigh Woods (10) (from 1789 until 1945) and several localities near Bath (10). The species is at its south-western limit in Britain in Somerset but whether this apparent decline is real or the species has been overlooked by recent observers is not clear.

EMPETRACEAE

EMPETRUM L.

Empetrum nigrum L. Crowberry

On high moorland. Rare but locally plentiful. On Haddon Hill (1), and on Exmoor between the Devonshire border and Dunkery Hill (1 & 2). Recorded also on Croydon Hill (2), 1952 (D. N. Williams) and once found on Quantock near Crowcombe (3) in 1852 (Miss Rosekelly fide T. Clark) but never seen there since.

PLUMBAGINACEAE

LIMONIUM Mill.

Limonium vulgare Mill. Common Sea Lavender

Statice limonium L.

In muddy places by the sea and in salt marshes. Local along the coast and up the tidal estuaries from Lilstock (2) to Clevedon (9). Formerly also at Portishead (9) but long since gone from there. Erroneously included in some Steep Holm lists and attributed to Banks' and Lightfoot's visit in 1773 but there is little doubt that the plant seen by them "on rocks on the southern side" was the next species which was not separated by British botanists until 1829.

Limonium binervosum (G. E. Sm.) C. E. Salmon Rock Sea Lavender

Statice auriculaefolia auct.

On rocks and sand near the sea. Rare and local. On the sea wall at Hinkley Point (2), 1975 (HWB), on fore dunes at Berrow (9) where it was first seen in 1906 (C. E. Moss), and on the rocks of Steep Holm, Brean Down, Weston-super-Mare and Sand Point in district (9). Also recorded from Stert Island (2) and a sea wall at Burnham (8) but these seem to have been impermanent occurrences as opposed to its stations on rocky and rather inaccessible cliffs where it has had a long history.

ARMERIA Willd.

Armeria maritima (Mill.) Willd. Thrift

Statice maritima Mill.

On cliff tops and in the drier parts of salt marshes. A garden escape inland. Formerly common near the sea but now local and often very sparse where it occurs. On cliff tops near Minehead (2), Steep Holm (very little), Brean Down, Sand Point and Clevedon (9). In salt marshes at Porlock (2), by the Parrett and Brue estuaries (8), at Uphill and from Wick St Lawrence to Clevedon (9). Also at Portbury (9) but probably eradicated by the building of the Royal Portbury Dock.

PRIMULACEAE

PRIMULA L.

Primula veris L. Cowslip

Dry grassland. Still common in the eastern half of the county on calcareous soils but it has suffered from the ploughing and re-seeding of old pastures. Very rare west of Taunton and the Quantocks. Not recorded in district 1.

× **vulgaris** (*P.* × *tommasinii* Gren. & Godron; *P.* × *variabilis* Goupil non Bast.).

Not infrequent in the eastern half of the county where the parents meet in meadows and on wood borders. Very rare in the west and only recorded recently near Minehead and in the Roadwater valley (2) (CJG).

Primula vulgaris Huds. Primrose

P. acaulis (L.) Hill

In woods and on hedgebanks. Very common except near large towns, where it has been diminished by digging up, and rather scarce in the Levels.

PRIMULACEAE

Hottonia L.

Hottonia palustris L. Water Violet

In marsh ditches and occasionally in pools. Locally plentiful in the Levels of districts 3, 4, 5, 8 and 9 south of Mendip. Outside this area it has only been seen recently at Nailsea Moor (9), 1976 (ESS), and at Tor Hole, Chewton Mendip (10), 1974 (JA), though it was reported in ditches at Puxton (9) in 1956 (ER). Its rarity in the lowlands north of Mendip is surprising.

Cyclamen L.

Cyclamen hederifolium Ait. Cyclamen

Naturalized in private grounds and sometimes escaping.

Lysimachia L.

Lysimachia nemorum L. Yellow Pimpernel

In woods and on shady hedgebanks. Common in the west and south and in wooded areas of the north and east but absent from most of the lowlands.

Lysimachia nummularia L. Creeping Jenny

Banks of marsh ditches, streams and rivers and in marshes. Very common throughout the lowlands and frequent in the eastern half of the county. Rare west of Taunton and probably always introduced there.

Lysimachia vulgaris L. Yellow Loosestrife

By rivers and streams, in marshy places and in damp woods. Frequent in the marshlands of districts 5, 8 and 9 south of Mendip. Rare elsewhere though there are scattered records in all districts, none recent in district 2.

Lysimachia punctata L. Dotted Loosestrife

A widespread garden escape.

†**Lysimachia thyrsiflora** (L.) DC Tufted Loosestrife

Introduced by Sole by the Avon near Twerton (10) in about 1780 it seems to have flourished for a good many years but has now long been extinct.

Monkshood, *Aconitum napellus.* Milverton
Photo: P. WAKELY

Mousetail, *Myosurus minimus.* West Sedgemoor
Photo: P. WAKELY

Cheddar Pink, *Dianthus gratianopolitanus.*
Photo: P. Wakely

Hairy Mallow, *Althaea hirsuta.* Aller Photo: P. Wakely

Purple Gromwell, *Buglossoides purpurocaerulea.* Rodney Stoke
Photo: P. WAKELY

Goldilocks, *Aster linosyris.* Photo: P. WAKELY

Bath Asparagus, *Ornithogalum pyrenaicum*.
Monkton Combe
Photo: P. Wakely

Spring Snowflake, *Leucojum vernum*.
Photo: P. Wakely

ANAGALLIS L.

Anagallis tenella (L.) L. — Bog Pimpernel

In boggy places especially round springheads and on acid soils. Common on Exmoor, the Quantock and Blackdown Hills and on Chard Common (6). In a few places on the peat moors (8), in the Gordano valley (9) and on the Mendip sandstone (9 & 10). Rare elsewhere. Not recorded in district 5 nor recently in 7.

Anagallis arvensis L. — Scarlet Pimpernel

A weed of cornfields and other cultivated land and sometimes in waste places. Very common throughout the county except on Exmoor where it is rare. Blue- and pink-flowered varieties sometimes occur.

Anagallis foemina Mill. — Blue Pimpernel

A. arvensis ssp. *foemina* (Mill.) Schinz & Thell.

Much less frequent than the blue-flowered variety of *A. arvensis*. Recorded recently only at Kingsdon (5), Charlton Mackrell (5) (where it was first collected by Gapper in 1822), and Nettlebridge, Lullington and Hinton Charterhouse in district 10.

†**Anagallis minima** (L.) E. H. L. Krause — Chaffweed

Centunculus minimus L.

Damp sandy tracks and woodland rides. Extinct or perhaps overlooked. Last recorded in 1951 at Winterhead near Shipham (9) by Miss E. Rawlins. Earlier records were near East Anstey (1), Luccombe, Wootton Courtenay and Alcombe (2), Penselwood (7), Nettlebridge (10) and in Leigh Woods (10).

GLAUX L.

Glaux maritima L. — Sea Milkwort

In muddy places and salt marshes by the sea and up the tidal rivers. Porlock (2), Carhampton (2) and locally common between Stolford (2) and the Avon estuary (10).

PRIMULACEAE

SAMOLUS L.

Samolus valerandi L. Brookweed

On ditchbanks and in marshy places where the water is alkaline. Frequent throughout the lowlands and near the sea in districts 2, 3, 4, 5, 8 and 9. Rare elsewhere and only seen recently near Milverton, Pitminster and West Hatch in district 3 and Broadway and West Chinnock in district 4. Formerly recorded further inland in districts 7 and 10. Babington (1834) said it was frequent near Bath (10) but in 1866 Jenyns said it was decidedly rare and it has not been seen in the area for over a century now.

BUDDLEJACEAE

BUDDLEJA L.

Buddleja davidii Franch. Butterfly Bush

Old walls, disused quarries and waste places. A shrub from China introduced to Britain about 1890 but not noticed as established in the wild in many places before 1930. The first published record for Somerset was in 1932 when Mrs. C. I. Sandwith noted it as established by the Avon under Leigh Woods (10). During the 1939-45 War it became abundant on the bombed sites of Bristol and Bath. It is now widespread, particularly in the limestone areas, in all districts but is still rare in the west and south.

OLEACEAE

FRAXINUS L.

Fraxinus excelsior L. Ash

Woods and hedges. Very common throughout the county.

SYRINGA L.

Syringa vulgaris L. Lilac

Hedges. Usually planted and sometimes marking the sites of demolished cottages. Frequent in the eastern half of the county. Rare in the west.

LIGUSTRUM L.

Ligustrum vulgare L. Wild Privet

Hedges, woods and downland scrub. Very common on calcareous soils. Rare in the west and on acid soils except as a planted hedgerow shrub. Plentiful on Steep Holm where it has been known since 1625.

Ligustrum ovalifolium Hassk. Garden Privet

Frequently planted in garden hedges.

APOCYNACEAE

VINCA L.

Vinca minor L. Lesser Periwinkle

On hedgebanks and in woods. Originally introduced. Usually an obvious escape from cultivation but sometimes well naturalized. Frequent except in the lowlands and the wilder parts of Exmoor.

Vinca major L. Greater Periwinkle

Hedgebanks and waste places. A frequent garden escape in all districts.

GENTIANACEAE

CENTAURIUM Hill

Centaurium pulchellum (Sw.) Druce Lesser Centaury

Erythraea pulchella (Sw.) Fr.; *E. ramosissima* (Vill.) Pers.

In calcareous grassland and dune slacks. Rather rare. Recently recorded on the Lias between Blue Anchor and Watchet (2), between West Hatch (3) and Bickenhall (4), around Somerton (5) and on the Polden Hills (5 & 8); on Oolite at Hardington (4), Maperton (5), Stoke Trister (7), Bruton (8) and between Norton St Philip and Bath (10). Formerly on the limestone west of Bristol (9 & 10) but not seen there since 1915. The only recent dune-slack records are from Berrow (9) though formerly it was recorded as far north as Weston-super-Mare (9).

Centaurium erythraea Rafn Common Centaury

C. minus auct.; *Erythraea centaurium* auct.

In dry grassland, open woods and on sand dunes. Common except on Exmoor and in the Levels. Recorded in all districts.

× **pulchellum.** Recorded from the coast of North Somerset. Although the parent species frequently grow together the hybrid has not been noticed anywhere else.

GENTIANACEAE

Centaurium capitatum (Willd.) Borbás Tufted Centaury

C. erythraea var. *capitatum* (Willd.) Melderis

This very rare little plant was first found on the southern slope of Crook Peak (9) in 1938 by F. K. Makins. It is still to be found there and in 1973 it was recorded (JA) on Wavering Down a little further east on Mendip. The main difference between this and *C. erythraea* is in the insertion of the stamens at the base, rather than near the top, of the corolla-tube, and some authorities do not consider it a distinct species.

(**Centaurium littorale** (D. Turner) Gilmour (*Erythraea littoralis* (D. Turner) Fr.) was recorded from Brean Down (9) by Collins (1837) but this has been considered an error by later botanists.)

BLACKSTONIA Huds.

Blackstonia perfoliata (L.) Huds. Yellow-wort

Grassland on calcareous soils. Common on the limestone throughout the county. Formerly common on the Berrow dunes (9) also but not recorded there recently. Not in district 1.

GENTIANELLA Borkh.

†**Gentianella campestris** (L.) Börner Field Gentian

Gentiana campestris L.

Long known near the north coast across the Devonshire border it was not located in Somerset until 1920 when N. G. Hadden found it in fields in three separate places between Oare and Porlock (2). It was seen in two of these by Hadden in 1940 but has not been reported since. Older records for Kilmington (7) (Wilts) and near Bath (10) were errors for *G. amarella.*

(**Gentianella germanica** (Willd.) E. F. Warb was credited to V.C.5 in Druce's *Comital Flora* (1932) but no record supporting this can be traced. The Rev E. F. Linton found it in 1893 on the chalk of Mere Down in Wiltshire just beyond the border of V.C.6.

Gentianella amarella (L.) Börner Autumn Gentian

Gentiana amarella L.

Grassland on calcareous soils. Frequent on the limestone but apparently less so than formerly. Not recorded in districts 1 or 6.

Gentianella anglica (Pugsl.) E. F. Warb. Early Gentian

Gentiana amarella var. *praecox* Towns.

Calcareous grassland. Very rare and only seen recently at Compton Dundon (5), Long Knoll (7) (Wilts) and Butleigh Hill(8). Formerly recorded also on White Sheet Downs (7) (Wilts), Bathford Hill and Lansdown near Bath (10).

MENYANTHACEAE

MENYANTHES L.

Menyanthes trifoliata L. Bogbean

In acid marshes, bogs and shallow pools. Frequent on Exmoor in district 1 (but surprisingly rare in district 2) and on the Blackdown Hills (4 & 6). Rare and decreasing elsewhere. Only recorded recently on the peat moors (8) and in pools at Priddy, Charterhouse, Churchill and Backwell in district 9. Sometimes introduced in ornamental ponds and lakes.

NYMPHOIDES Séguier

Nymphoides peltata (S. G. Gmel.) Kuntze Fringed Water-lily

A rare escape in Somerset. First recorded in a mill leat at Cannington (3), 1915 (H. Slater), but soon afterwards cleared out. Recent records are in the canal at North Newton (3), 1975 (R. K. Banfield), in the Tone at Creech St Michael (3), 1976 (F. R. Gomm) and in a pond at Paulton (10), 1979 (DG).

BORAGINACEAE

CYNOGLOSSUM L.

Cynoglossum officinale L. Hound's-tongue

In thin turf on limestone downs and on sand dunes. Locally frequent near the sea about Minehead (2) and between Burnham (8) and Wick St Lawrence (9), and on Mendip west of Banwell and Loxton (9). Decreasing elsewhere and only seen recently near Yeovil (4), on the Lias around Somerton (5), at Cadbury Castle (5), Goblin Combe (9) and on the downs round Bath (10). Formerly more widespread particularly in the north of the county.

BORAGINACEAE

OMPHALODES Mill.

Omphalodes verna Moench — Blue-eyed Mary

Garden escape.

ASPERUGO L.

†**Asperugo procumbens** L. — Madwort

This was known as a cornfield weed at Twerton near Bath (10) for over a hundred years but it has not been seen there since 1916. It has occurred as a casual on waste ground elsewhere in districts 9 and 10 but not since 1928.

SYMPHYTUM L.

Symphytum officinale L. — Common Comfrey

By ditches, streams, in marshes and damp meadows. Very common except on Exmoor. The purple-flowered variety (var. **purpureum** Pers.) is almost as common as the white-flowered var. **ochroleucum** DC., but perhaps in drier places.

Symphytum asperum Lepech. — Rough Comfrey

An introduced fodder plant which may occur in a few places in the county as an escape from cultivation but it has not been reliably reported recently.

× **officinale** (*S.* × *uplandicum* Nyman; *S. peregrinum* auct.) (Russian Comfrey). A very variable plant, originally introduced for fodder but which may also have arisen spontaneously in Britain. Frequent on road-sides, but its exact distribution in the county is not known because of confusion with *S. officinale* var. *purpureum*. Most reports of *S. asperum* are this hybrid. Its variability is probably due to back-crossing with *S. officinale*.

Symphytum orientale L. — White Comfrey

An uncommon garden escape.

Symphytum ibiricum Steven — Creeping Comfrey

S. grandiflorum auct.

A garden escape which is naturalized and persistent in a few places.

BORAGO L.

Borago officinalis L. Borage

A frequent garden escape.

PENTAGLOTTIS Tausch

Pentaglottis sempervirens (L.) Tausch Green Alkanet

Anchusa sempervirens L.

On roadsides, by hedges and in waste places. Originally introduced but now well established and frequent in all districts particularly in the west.

ANCHUSA L.

Anchusa azurea Mill.

A. italica Retz.

Garden escape.

Anchusa arvensis (L.) Bieb. Bugloss

Lycopsis arvensis L.

In sandy places near the sea, and a decreasing weed in cultivated land and waste places. Fairly common on the coast near Minehead (2) and between Burnham (8) and Kewstoke (9). Rather rare inland and not recorded in districts 1, 6 or 7.

PULMONARIA L.

(**Pulmonaria longifolia** (Bast.) Bor. (*P. angustifolia* auct.) was credited to V.C.6 in Watson's *Topographical Botany* (1873-74) on the strength of a specimen in Dr H. O. Stephen's herbarium but there is no evidence that it was collected in North Somerset.)

Pulmonaria officinalis L. Lungwort

Garden escape.

BRUNNERA Stev.

Brunnera macrophylla (Adams) I. M. Johnston

Garden escape.

BORAGINACEAE

Myosotis L.

Myosotis scorpioides L. Water Forget-me-not

M. palustris (L.) Hill

By rivers, streams and ponds and in marshy places. Common over much of the county but scarce in the west.

Myosotis secunda A. Murr. Creeping Forget-me-not

M. repens auct.

By springs, ditches and streams and in wet boggy ground on acid soils. Common on Exmoor and on the Brendon, Quantock and Blackdown Hills and on Chard Common (6). Elsewhere it has only been seen recently at a few spots on the flanks of the Mendip sandstone (8 & 9) and near Witham (10). White (1912) gives a number of localities on the moors of north Somerset but the plant has not been recorded since anywhere in the area.

Myosotis laxa Lehm. ssp. **caespitosa** (C. F. Schultz) Hyl. ex Nordh. Tufted Forget-me-not

M. caespitosa C. F. Schultz

By ditches, streams and ponds. Common, especially in the lowlands, over much of the county but scarce in the west.

Myosotis sylvatica Hoffm. Wood Forget-me-not

An uncommon garden escape in Somerset. Large-flowered varieties of *M. arvensis* are sometimes mistaken for this.

Myosotis arvensis (L.) Hill Field Forget-me-not

A weed of cultivated ground, roadsides and waste places. Very common throughout the county. A larger-flowered cultivated variety is a frequent garden escape.

Myosotis discolor Pers. Changing Forget-me-not

M. versicolor (Pers.) Sm.

On dry banks, wall tops, woodland tracks and sandy ground. Rather common and recorded in all districts.

Myosotis ramosissima Rochel Early Forget-me-not

M. hispida Schlecht; *M. collina* auct.

In dry grassland on calcareous soils and on sand dunes. Frequent in these habitats, rare elsewhere. Not recorded in district 1.

LITHOSPERMUM L.

Lithospermum officinale L. Common Gromwell

In woods, downland scrub and hedges on calcareous soils. Frequent on the Carboniferous Limestone, Oolite and Lias in the eastern half of the county. Rare in the west and only seen recently at Cleeve Hill (2) and near Milverton, Cannington and Stoke St Mary in district 3. Not recorded in districts 1 or 6.

BUGLOSSOIDES Moench

Buglossoides purpurocaerulea (L.) I. M. Johnston Purple Gromwell

Lithospermum purpurocaeruleum L.

On and near wood borders and in scrub on limestone. Rare and local. It is probably more frequent in the Mendip woods between Hutton and Rodney Stoke (9) than anywhere else in Britain. Elsewhere it has only been seen recently at Fivehead (3), Aller (5), Brent Knoll (8), Goblin Combe (9) and Weston-in-Gordano (9), in all of which localities it has a long history. The first British record, by Ray (1690), "not far from Taunton" probably refers to Fivehead where it was also recorded by the Rev C. Parish in 1849. Formerly recorded in several places on the Polden Hills between Somerton and Puriton (5 & 8) but not noted there since 1943; it may well have been overlooked as it is shy in flowering and not conspicuous at other times.

Buglossoides arvensis (L.) I. M. Johnston Field Gromwell

Lithospermum arvense L.

A weed of cornfields and waste places. Very rare and decreasing. Only seen recently as a weed at Porlock (2), 1965 (NGH), in a disused railway yard at Pitcombe (8), 1964 (CAH), and in several localities in the north-east corner of the county round Bath (10). Formerly much more wide-spread in districts 3, 4, 5, 9 and 10, the decline seems to have taken place since 1946 no doubt due to the increased use of weed killers.

BORAGINACEAE

ECHIUM L.

Echium vulgare L. Viper's Bugloss

Dry roadside and railway banks, old walls and sandy waste ground. Rather rare. Seen recently only at Minehead Warren (2), Broomfield (3), Lydford and Burnham (8), Axbridge, Bleadon, Worle, Yatton and Tickenham (9), Clutton and in the Mells area (10). Formerly recorded also in districts 4 and 5.

Echium plantagineum L. Purple Viper's Bugloss

E. lycopsis L. pro parte

A casual or garden escape which appeared in several places by a short stretch of the new M5 motorway near Clevedon (9) in 1973 and 1974 (JA).

CONVOLVULACEAE

CONVOLVULUS L.

Convolvulus arvensis L. Field Bindweed

A troublesome weed of cultivated land and waste places. Very common everywhere in the county except in the wilder parts of Exmoor.

CALYSTEGIA R. Br.

Calystegia sepium (L.) R. Br. Hedge Bindweed

Convolvulus sepium L.

Hedges, willow holts, river banks and a weed in gardens. Very common in all districts.

× **silvatica** (*C.* × *lucana* (Ten.) G. Don). Although said to be frequent where the parent species are common, as in many parts of Somerset, this hybrid has not been positively identified in the county yet.

Calystegia pulchra Brummitt & Heywood Hairy Bindweed

C. dahurica auct.

An uncommon garden escape.

Calystegia silvatica (Kit.) Griseb. Large Bindweed

C. sylvestris (Willd.) Roem. & Schult.

Hedges, railway banks and waste places. Originally a garden escape but now common throughout the county especially near towns and villages. Not recorded in Somerset until 1938 [near Bath (10) (Miss A. E. White)] although it must have been widespread long before this.

Calystegia soldanella (L.) R. Br. Sea Bindweed

Convolvulus soldanella L.

On sandy shores. Rare and decreasing on the coast between Burnham (8) and Kewstoke (9). Formerly also at Porlock Weir and Steart (2) but not seen in either locality for many years.

IPOMOEA L.

Ipomoea hederacea (L.) Jacq.

A rare casual found in a farmyard at Gare Hill (10) in 1975 (PT).

CUSCUTA L.

Cuscuta europaea L. Greater Dodder

On river banks, parasitic on Nettles and sometimes on other plants. Rare and local. Said by Gerard (1597) to be "very plentiful on Nettles in Somersetshire" it still occurs at intervals on the banks of the Avon between Keynsham (10) and the Wiltshire border. It was also found by the Yeo near Muchelney (4) in 1963 (MHP). Older records are at Charlton (5), presumably by the Cary, in 1822 (Gapper) and at Butleigh (8) in 1848 (T. Clark). Records by J. C. Collins for hills near Minehead and on the Quantocks were undoubtedly errors for *C. epithymum.*

†**Cuscuta epilinum** Weihe Flax Dodder

Parasitic on Flax, which was formerly cultivated in the north of the county, but not recorded since the middle of the nineteenth century.

Cuscuta epithymum (L.) L. Dodder

including *C. trifolii* Bab.

A parasite on various shrubs and herbs. Frequent on Gorse and Ling on the acid soils of the Quantocks and the northern part of Exmoor but strangely absent from the southern part in district 1. Formerly on the Blackdown Hills too but not recorded there since 1920. On calcareous soils the host plants are small *Leguminosae,* thyme, carrot, bedstraws, etc. Formerly a pest in clover fields and quite widespread in the county but now rare and recently recorded only at Cleeve Hill and Lilstock (2), Thurlbear (3), Aller and Charlton Mackrell (5), Butleigh (8), Clevedon (9) and Wellow (10).

HYDROPHYLLACEAE

PHACELIA Juss.

Phacelia tanacetifolia Benth.

An uncommon garden escape found in new-sown grass at Hillfarrance (3) in 1967 (D. Tucker).

SOLANACEAE

NICANDRA Adans.

Nicandra physalodes (L.) Gaertn. Apple of Peru

Garden escape. Sometimes called Shoo-fly Plant from its supposed ability to rid greenhouses of the pest Whitefly. First recorded in Somerset at Ashton Gate (10), 1918 (Misses Cobbe) and since at many other places.

LYCIUM L.

Lycium barbarum L. Duke of Argyll's Tea Plant

L. halimifolium Mill.

Planted in hedges and sometimes an established escape. Widespread.

Lycium chinense Mill.

Less commonly planted than *L. barbarum* but now thoroughly established on sandy ground near the coast especially between Burnham (8) and Kewstoke (9).

ATROPA L.

Atropa bella-donna L. Deadly Nightshade

In woods and waste places. Very rare. Probably native on the coast near Lilstock (2), in Leigh Woods, near Farleigh Hungerford and on Hampton Down near Bath (10), in all of which places it has a long history. A rare casual or escape from cultivation elsewhere, seen recently at Bridgwater (3), Charlton Adam (5), Wells (8) and at several spots near Bath (10).

HYOSCYAMUS L.

Hyoscyamus niger L. Henbane

In sandy places near the sea and a weed of farms and waste places inland. Rare and sporadic in appearance. Native on the coast in a few places between Minehead (2) and Kewstoke (9). More plentiful and constant on Steep Holm, where it has been known since 1826 (W. Withering fil.), than anywhere else. Formerly as a casual weed inland but nowhere persistent except on the banks of the Yeo between Long Load and Ilchester (4 & 5). The only other recent records were at Charlton Horethorne (5), and at Frome, Ashwick, Compton Dando and Brislington, all in district 10.

PHYSALIS L.

Physalis peruviana L.

Casual.

SOLANUM L.

Solanum dulcamara L. Bittersweet

In hedges, ditches, woods and on waste ground. Very common throughout the county except in the extreme west where it is rare.

Solanum nigrum L. Black Nightshade

A weed of gardens, arable land and waste places. Common except on Exmoor and the hills. Not recorded in district 1.

Solanum tuberosum L. Potato

A frequent relic of cultivation in farmyards and on rubbish tips.

Solanum rostratum Dunal

S. cornutum auct.

A rare casual probably introduced with grass seed, first recorded in the county at St Anne's, Brislington (10), 1922 (HJG) and seen more recently at Yeovil (4) and Frome (10).

LYCOPERSICON Mill.

Lycopersicum esculentum Mill. Tomato

Generally cultivated and widespread as an escape or casual on rubbish tips, by railways, on sewage farms and in other waste places.

SOLANACEAE

DATURA L.

Datura stramonium L. Thorn-apple

A casual of gardens and waste land. Rare but widespread and sometimes persistent as the seeds can lie dormant for years, plants springing up again when the ground is disturbed. Highly poisonous. Periodically the subject of alarmist press reports about the "deadly American Thorn-apple" as if it were a newcomer. The earliest Somerset record "Waste places about Bath" (where it still appears from time to time) by Sole dates from 1801. Not yet recorded in districts 1, 6 or 7.

SCROPHULARIACEAE

VERBASCUM L.

Verbascum thapsus L. Great Mullein

In grassy and waste places. Common and widely distributed, particularly on calcareous and sandy soils.

Verbascum phlomoides L. Orange Mullein

An uncommon garden escape first recorded in the county in 1930 at Clevedon (9) (Miss M. A. G. Livett).

Verbascum lychnitis L. White Mullein

Dry grassland, woods and waste places. Very rare. The yellow-flowered variety is native about Bossington and Selworthy (2). Elsewhere it has only been recorded recently as a casual in two spots near Bath (10). There are old records for the species near Taunton (3), Banwell and Worle (9), Beckington, Bathford and Lyncombe, Bath (10). These were all probably casual occurrences.

× **thapsus** (*V.* × *thapsi* L.) was recorded near Bossington (2) in 1924 and 1930.

Verbascum nigrum L. Dark Mullein

Roadsides and dry grassland on calcareous soils. Only a rare casual in Somerset although frequent on the Oolite of Gloucestershire and North Wiltshire not far across the border. The only recent record was as a weed in nursery gardens on Bannerdown (10), 1976 (TC). There are older records as a casual in districts 3, 8, 9 and 10.

Verbascum blattaria L. Moth Mullein

A rare casual. The only locality in which it seems to be firmly established is near Winscombe (9) where it has been known for well over a century. There are other recent records in districts 3, 7, 9 and 10 and old ones in 2, 4 and 5 as well but seldom twice in the same place.

Verbascum virgatum Stokes Twiggy Mullein

A rare casual, similar in occurrence and distribution to the last species. It too has a long history in the Winscombe area where it is still to be found. Other recent records are in districts 4, 5, 8, 9 and 10 and old ones in 2 and 3 also.

MISOPATES Raf.

Misopates orontium (L.) Raf. Lesser Snapdragon

Antirrhinum orontium L.

Always very rare as a weed of cultivated land in Somerset, the only recent record is near Porlock (2), 1977 (CJG). It also occurred in 1977 as a bird-seed alien with many other species at Kilmington (7) (Wilts) (PT). Former records were in districts 3, 8, 9 and 10, but only two of these since 1920.

ANTIRRHINUM L.

Antirrhinum majus L. Snapdragon

A garden escape frequently naturalized on old walls and in waste places. Recorded in all districts except 6.

LINARIA Mill.

Linaria purpurea (L.) Mill. Purple Toadflax

A garden escape which is becoming increasingly common and well naturalized on walls and in waste places in all districts.

Linaria repens (L.) Mill. Pale Toadflax

By railway tracks, on walls and in waste places. Spread along the railways from its native stations on the chalk further east and a garden escape in other places. Recorded recently by the railway at Somerton and Castle Cary (5), Evercreech (8), Clutton and Wellow (10). An escape on walls in districts 1, 2, 5, 8 and 10. Long established on walls at Locking (9) but not seen there since 1918 although reported as a troublesome weed in a garden at Banwell not far away.

× **vulgaris** (*L.* × *sepium* Allm.). Found in 1969 on the disused railway track near Clutton (10), with the parents.

Linaria vulgaris Mill. Common Toadflax

On hedgebanks, field borders, railway banks and in waste places. Common and recorded in all districts.

Linaria genistifolia (L.) Mill. ssp. **dalmatica** (L.) Maire & Petitmengin

L. dalmatica (L.) Mill.

Escaped from nursery gardens on to railway banks at Langport (5) in 1916 (ESM) and still established there and at Curry Rivel (3) a mile further west. Reported as an impermanent escape in one or two other places.

CHAENORHINUM (DC.) Reichb.

Chaenorhinum minus (L.) Lange Small Toadflax

Linaria viscida Moench; *L. minor* (L.) Desf.

On railway tracks and a weed in cultivated and waste ground. Rather common about railways but less so in other places. Recorded in all districts.

KICKXIA Dumort.

Kickxia spuria (L.) Dumort. Round-leaved Fluellen

Linaria spuria (L.) Mill.

A weed of cornfields and other cultivated land. Frequent on the Lias of districts 2, 3, 4 and 5 and the Oolite of district 10. Rather rare in districts 7, 8 and 9.

Kickxia elatine (L.) Dumort. Sharp-leaved Fluellen

Linaria elatine (L.) Mill.

A weed of cornfields and arable land. Formerly more widespread and more frequent than *K. spuria* particularly in the north of the county but decreasing and now rarer. The two species often grow together and their distribution is similar except that *K. elatine* has not been recorded in district 7 and there has been one record in district 6, at Whitestaunton, 1920 (WW).

CYMBALARIA Hill

Cymbalaria muralis Gaertn., Mey & Scherb. Ivy-leaved Toadflax

Linaria cymbalaria (L.) Mill.

Originally introduced but long and thoroughly naturalized on walls. Very common throughout the county. A white-flowered variety is quite frequent, especially in the Bath area (10).

Cymbalaria pallida (Ten.) Wettst.

A rare alien established on a wall at Taunton (3), 1964 (D. W. Livsey).

SCROPHULARIA L.

Scrophularia nodosa L. Common Figwort

Woods and hedgebanks. Very common over most of the county but rather sparse on the Levels.

Scrophularia auriculata L. Water Figwort

S. aquatica auct.; *S. oblongifolia* Lois.

By ditches and streams in damp places. Very common in all districts except on the highest parts of Exmoor.

MIMULUS L.

Mimulus guttatus DC. Monkey Flower

M. langsdorffii Donn ex Greene; *M. luteus* auct.

By streams and rivers and in marshy places. Originally a garden escape but it has spread along streams in the west of the county and has become quite common in districts 1, 2, 3 and 6. Elsewhere it is still rare and confined to a few localities in districts 4, 7, 8, 9 and 10.

× **luteus.** First positively identified in the county by R. H. Roberts from specimens collected by H. W. Boon, it now appears that this hybrid is frequent in the streams of Exmoor and the Quantocks, and that many of the past records for *M. guttatus* really belong here. In and by the Barle near Simonsbath (1) the true *M. guttatus* is sparse but the hybrid, less pubescent and with variable blotching on the corollas, is plentiful.

SCROPHULARIACEAE

Mimulus luteus L. Blood-drop Emlets

A rare garden escape, sometimes reported in error for the hybrid with *M. guttatus*. Apparently not so hardy as either *M. guttatus* or the hybrid which would explain its rarity in the wild.

Mimulus moschatus Dougl. ex Lindl. Musk

By streams in boggy and marshy places. A rare garden escape first recorded in the county on Ashcott Heath (8), 1909 (F. Samson), but no longer to be found there. It has become thoroughly naturalized since 1911 (H. Corder) in Holford Combe (2), and since 1918 (WW) at Wambrook (6) (V.C.9). Reported recently also at Nettlecombe (2), Clatworthy Reservoir (3), 1975 (Miss M. Davidson) and Churchingford (6) (V.C.3), 1975 (MGMcF). Recorded formerly in districts 1 and 7 but not seen since in either.

LIMOSELLA L.

†**Limosella aquatica** . Mudwort

Formerly in muddy places. Recorded by Coleman near Dunster (2) in 1848 and by Sole (1791) near Highbridge (8) but not seen anywhere in the county since.

SIBTHORPIA L.

Sibthorpia europaea L. Cornish Moneywort

On moist banks by streams and ditches. Rare. On the borders of Exmoor and the Brendons (1, 2 & 3) and round the northern end of the Quantocks (2 & 3). Once reported from the Blackdowns, below the Wellington Monument (3), 1952 (A. J. Dodd), but not seen there since.

ERINUS L.

Erinus alpinus L. Fairy Foxglove

Garden escape.

DIGITALIS L.

Digitalis purpurea L. Foxglove

On hedgebanks, in open woods and on heaths. Very common on acid soils particularly on Exmoor and the other hills of the west and south. Rare in the limestone areas and usually only a garden escape there.

VERONICA L.

Veronica beccabunga L. Brooklime

By streams and ditches, in marshes and on the muddy edges of ponds. Very common throughout the county.

Veronica anagallis-aquatica L. Blue Water-speedwell

By streams and ditches and in other wet places. Frequent in the eastern half of the county but very rare west of Taunton (3). Not recorded recently in districts 1 or 2.

Veronica catenata Pennell Pink Water-speedwell

V. aquatica Bernh.

In ditches, ponds and slow-moving streams. Very common throughout the Levels. Rather rare elsewhere and not recorded in districts 1 or 6. This species was not distinguished from *V. anagallis-aquatica* by Somerset botanists until the 1930s and many of the older records of the latter are of this. The two species sometimes grow together but this is the commoner in the marshlands and *V. anagallis-aquatica* away from them.

Veronica scutellata L. Marsh Speedwell

In marshes and bogs, usually on acid soils. Frequent on Exmoor, the Brendon and Blackdown Hills and the peat moors (8). Very rare elsewhere and only recorded recently at Kingsdon (5), 1976 (R & IR), Kenn Moor (9), 1963 (RMB), Weston-in-Gordano (9), 1966 (CHC), at a few spots on the Mendip sandstone (9 & 10), Witham, 1975 (CAH), Kilmersdon, 1980 (DG), and Bathford (10), 1962 (R & IR). Not seen recently in district 7.

Veronica officinalis L. Heath Speedwell

On dry banks, heaths and acid grassland. Common in the west and south, on the peat moors (8) and on Mendip, both on sandstone and on limestone heath where locally acid conditions have developed by leaching. Frequent elsewhere in suitable places except in the lowlands.

Veronica montana L. Wood Speedwell

In woods and shady lanes. Common over most of the county except where there is no suitable habitat, as on the Levels.

SCROPHULARIACEAE

Veronica chamaedrys L. Germander Speedwell

Hedgebanks, roadside verges, open woods and grassland. Very common throughout the county.

Veronica spicata L. Spiked Speedwell

V. hybrida L.; *V. spicata* ssp. *hybrida* (L.) E. F. Warb.

On limestone rocks. Very rare. Known in the Avon Gorge since 1641 it was not generally believed to occur on the Somerset side until 1912 when it was found on rocks under Leigh Woods (10) by Miss I. M. Roper. Since then it has been found in two other localities in the woods (PJMN). In 1970 the existence of a herbarium book of 1779, compiled by Dr A. Broughton, came to light. In the book was a manuscript list of species with localities and this included *"Veronica hybrida"* from Leigh Woods, which becomes the earliest Somerset record. Plants from the Avon Gorge and Wales are larger than those from East Anglia and have been regarded as a separate species or subspecies but there seems little justification for this. This species was another of the rarities subjected to a transplant experiment in 1955 (see *Arabis stricta*).

Veronica serpyllifolia L. Thyme-leaved Speedwell

In damp grassland, on muddy tracks and in damp waste places. Very common in all districts.

Veronica peregrina L. American Speedwell

A weed of cultivation from America first recorded in Somerset in a garden at Bruton (8) in 1970 (CAH) and since reported as abundant in a nursery garden at Merriott (4), 1978 (R & IR).

Veronica arvensis L. Wall Speedwell

Wall tops, dry banks and a weed in cultivated land. Very common in all districts.

Veronica acinifolia L. French Speedwell

A European weed of cultivated ground rarely seen in Britain. Abundant in nursery gardens at Merriott (4), 1978 (R & IR).

Veronica hederifolia L. sensu lato Ivy-leaved Speedwell

Hedgebanks, woods and a weed in cultivated land. Very common in all districts. Dr M. Fischer has recently shown that the aggregate consists of five microspecies, two of which, **V. hederifolia** L. sens. strict., and **V. sublobata** M. Fischer (*V. hederifolia* ssp. *lucorum* (Klett & Richter) Hartl), occur in Britain. The former is found mainly in arable land whereas *V. sublobata* prefers shady, damp habitats, including woodland and gardens. The distributions of the two species in Somerset has not been worked out yet but *V. sublobata* seems the more common.

Veronica persica Poir. Common Field Speedwell

V. buxbaumii Ten., non Schmidt; *V. tournefortii* auct.

A weed of cultivated and waste land. Very common in all districts. First recorded in Britain in 1825 it spread very rapidly about the middle of the nineteenth century and is now thoroughly naturalized.

Veronica polita Fr. Grey Field Speedwell

V. didyma auct.

A weed of gardens and arable land. Common and widely distributed throughout the county but apparently decreasing.

Veronica agrestis L. Green Field Speedwell

A weed in cultivated land and waste places. Formerly common but now rather rare and still decreasing. Recorded from all districts except 6 but absent from large tracts of the county.

Veronica filiformis Sm. Slender Speedwell

Pastures, especially by streams and rivers, roadsides and lawns. Cultivated in Britain since 1808 it became popular as a rock-garden plant about 1920. First recorded as an aggressive weed in Somerset at Bicknoller (2), 1932 (T. A. Sprague), it had become established in more natural habitats by 1946 and from then on rapidly became more widespread. It is shown in the *B.S.B.I. Atlas* (1962) as recorded in about half the 10 km squares Somerset extends into; it has now been recorded in all except a few which include only small corners of the county.

Veronica crista-galli Stev.

This alien from south-west Asia is naturalized on hedgebanks and grassy roadsides in a small area at Batheaston (10) where it was first noticed by L. V. Lester-Garland in 1926. It was recorded again by Miss A. E. White in

1941 and R. D. Randall in 1977 though not seen apparently in intervening years. Two outlying patches about a mile from the main concentration were found in 1979, one in Bathford (10) (RDR) where the plant had also been seen in 1935 (JPMB), and the other still in Batheaston parish but nearer Bath (DG). This is at present the only known locality for the species in England though it is widespread in south-west Ireland and has also been found near Aberystwyth. It was naturalized in Sussex for over 80 years until the locality was destroyed by roadworks; it was believed to have escaped from the garden of H. Borrer. The Batheaston locality borders the grounds of a house which was once the home of C. E. Broome who may well have been responsible for its introduction.

PEDICULARIS L.

Pedicularis palustris L. Marsh Lousewort

In peaty bogs and springheads. Fairly common on Exmoor, the Brendon and Blackdown Hills and the peat moors (8). Rare elsewhere and only reported recently from near Wiveliscombe, Langford Budville, 1967 (CLH), and Enmore, 1971 (HWB), in district 3 and a few spots on Blackdown, Mendip (9). Not seen recently in district 10 although formerly recorded at Downhead, Frome East Woodlands and Claverton Down near Bath.

Pedicularis sylvatica L. Lousewort

Damp pastures and heaths on acid soils. Common on the moors and hills of the west and south, on the peat moors (8), and on the Mendip sandstone and limestone heath. Rather rare elsewhere and apparently decreasing but recorded in all districts.

RHINANTHUS L.

(**Rhinanthus angustifolius** C. C. Gmel. (*R. serotinus* (Schönh.) Oborny; *R. major* auct., non L.) Reported on the peat moors (8) by T. Clark and several later botanists including Marshall but the identification was doubted by other authorities. N. Y. Sandwith concluded in 1949 that Marshall's specimen at Kew was a late-flowering, robust form of *R. minor*. A specimen in the Taunton herbarium, collected by H. S. Thompson from Kenn Moor (9) in 1918, appears to be the same.)

Rhinanthus minor L. Yellow Rattle

In meadows and other grassland. Common throughout the county.

MELAMPYRUM L.

Melampyrum pratense L. Common Cow-wheat

In open woodland and on heaths. Common on Exmoor and the Quantocks and frequent elsewhere in the west and south (1, 2, 3, 4 & 6). Rare and decreasing elsewhere. The only recent records are at Brewham (8), Shipham, Congresbury and Weston-in-Gordano (9). Not recorded recently in districts 7 or 10 although there are many older records in the north of the county.

(**Melampyrum sylvaticum** L. (Small Cow-wheat). T. B. Flower's record of this in the woods at Hinton Abbey (10) in Babington (1839) is considered to have been an error for a small example of *M. pratense*.)

EUPHRASIA L.

Euphrasia officinalis L. sensu lato Eyebright

In grassland on downs, heaths and moors. Common as an aggregate throughout the county. Opinions on the division into segregates have changed considerably over the years and the distribution of the micro-species, as now accepted, requires further work.

Euphrasia micrantha Reichb. Slender Eyebright

E. gracilis (Fr.) Drej.

Acid heaths, often with *Calluna vulgaris*. Recorded as rare and very local on Exmoor and the Brendons (1 & 2) but it has not been positively identified since 1952 although probably still there.

†**Euphrasia scottica** Wettst. Slender Scottish Eyebright

In boggy springheads on moorland. Recorded from a few spots in the extreme west of Exmoor (1 & 2) but not seen there since 1918.

Euphrasia tetraquetra (Bréb.) Arrond. Broad-leaved Eyebright

E. occidentalis Wettst.

On grassy cliff tops and sand dunes and on limestone grassland further inland. Rather rare. At a few places near the coast between Porlock (2) and St Thomas Head (9), and on Mendip as far inland as Charterhouse (9). There are inland records by C. A. Howe, determined by P. F. Yeo, near Compton Dundon (5), South Cadbury (5) and Pitcombe (8).

SCROPHULARIACEAE

Euphrasia nemorosa (Pers.) Wallr. Common Eyebright

The most widespread of the microspecies. Common on calcareous soils it occurs less frequently on neutral and acid ones. Recorded in all districts. A strongly hairy variety of this was formerly referred to as **E. curta** (Fr.) Wettst., and has been recorded in the county.

Euphrasia confusa Pugsl. Little Kneeling Eyebright

E. minima auct.; *E. curta* var. *glabrescens* sensu White pro parte

In short turf on upland pastures. Frequent on Exmoor and the Brendons (1, 2 & 3), it also occurs in the south-east of the county on hilly pastures and downs between Somerton and the Wiltshire border (5, 7 & 8) and on the Mendips above Cheddar (9).

× **tetraquetra.** A specimen from Pitcombe (8) found by C. A. Howe in 1966 was identified as this hybrid by P. F. Yeo and it may be expected in other places, such as the Mendips above Cheddar, where the parent species grow together.

†**Euphrasia pseudokerneri** Pugsl. Chalk Hill Eyebright

A specimen from Bath dated 1830 in Babington's herbarium at Cambridge University was identified as this species by P. F. Yeo. Records under *E. kerneri* in White (1912) and Marshall (1914) were almost certainly forms of *E. nemorosa*.

Euphrasia arctica Lange ex Rostrup ssp. **borealis** (Townsend) Yeo Greater Eyebright

E. borealis (Towns.) Wettst.; *E. brevipila* auct.; *E. stricta* auct.

In rough grassland. Rather rare. Only recorded recently on the peat moors (8) though formerly near Simonsbath (1), Bossington (2) and in a few places on Mendip (9) also.

Euphrasia anglica Pugsl. English Sticky Eyebright

E. rostkoviana auct.; *E. fennica* auct.

In rough grassland and on grassy roadsides on acid soils. Common on Exmoor and the Brendons (1, 2 & 3). Frequent on the Quantocks, Blackdowns (3, 4 & 6) and the Greensand of the Wiltshire border (7, 8 & 10). It also occurs on the peat moors (8) and in a few places on the Mendip sandstone.

× **micrantha.** Found near Withypool (1) in 1952 by P. F. Yeo, both in a sterile and in a fertile form which might be recognized as a new and as yet unnamed microspecies (see *Watsonia* **3.** 253-269, 1956).

× **confusa** (*E. rechingeri* auct. non Wettst.) has been identified where the parents grow together as on Exmoor and the Brendons and on Mendip.

× **arctica.** Only on the peat moors (8). Specimens in the Cambridge University herbarium, collected there in 1922 by H. S. Thompson, have been identified as this hybrid by P. F. Yeo, as have later gatherings.

ODONTITES Ludw.

Odontites verna (Bellardi) Dumort. Red Bartsia

Bartsia odontites (L.) Huds.

Roadsides, cornfields and waste places. Common throughout the county. Most of our plants belong to ssp. **serotina** (Dumort.) Corb., with bracts not exceeding the flowers and widely spreading branches. Ssp. **verna** has been recorded in a few places but has not been identified recently.

OROBANCHACEAE

LATHRAEA L.

Lathraea squamaria L. Toothwort

A parasite on roots, mainly of Hazel but also on Wych Elm. Rare in South Somerset and only seen recently near Cutcombe (2), 1978 (CJG), Sherford and Pitminster (3), Chaffcombe (4) and Buckland St Mary (6). Rare also in districts 7 and 8 but frequent in some of the Mendip woods (9 & 10) and in the woodland near Bristol and around Bath (9 & 10).

Lathraea clandestina L. Purple Toothwort

An introduced plant parasitic on the roots of Willow and Poplar, reported in an overgrown shrubbery by a stream at Lower Weston, Bath (10) in 1954 (M. Gilwhite). In 1967 when the area was being cleared some plants were transferred to the roots of Willow in the Bath Botanic Garden, but it still survives in the original locality.

OROBANCHACEAE

OROBANCHE L.

†Orobanche purpurea Jacq. Yarrow Broomrape

A parasite on Yarrow. This very rare and decreasing species was recorded near Merridge (3) in 1869 by the Rev W. Tuckwell but has not been seen since.

Orobanche rapum-genistae Thuill. Greater Broomrape

O. major auct.

Parasitic on Broom and Gorse. Very rare and uncertain in its appearance. The only recent records are on Gorse near Dulverton (1), 1976 (LJM), and on Broom near Clutton (10), 1970 (PA & MH), and Kilmersdon (10), 1980 (DG). Previously it had been recorded at a few spots in districts 2, 3, 7 and 9 but had not been seen since 1907.

Orobanche elatior Sutton Knapweed Broomrape

A parasite on Greater Knapweed. Very rare and perhaps extinct in the county. Last recorded near Taunton (3), 1953 (F. Clarke). There are older records from Cheddar, Clevedon and Portishead (9). Reports from Brean Down (9) by Collins and St Brody were considered to be errors for *O. hederae* by Murray and White. Among several records for the Bath area in Babington (1834) is one by Dr H. Gibbes from the spur of the Cotswolds above Weston (10); it has been reported there several times since, most recently in 1919, by T. H. Green who said it appeared yearly. It is believed he saw it in succeeding years, perhaps as recently as 1941.

Orobanche minor Sm. Common Broomrape

Parasitic on many species, particularly clover. Frequent and widespread. Recorded in all districts except 1 and not recently in 7.

Orobanche loricata Reichenb. Oxtongue Broomrape

O. picridis F. W. Schultz ex Koch

Parasitic on *Crepis* and *Picris*. Very rare. First recorded in Somerset in 1894 at Brislington (10) (D. Fry det. Prof. G. Beck). There were no confirmed later sightings until 1969 when D. J. Hambler identified as this a specimen from a colony of ten plants found by P. Macpherson on Brean Down (9) (**TTN**).

Orobanche hederae Duby Ivy Broomrape

Parasitic on Ivy, especially near the sea. Rare near the coast between Porlock and Minehead (2). Quite frequent from Brean Down to Clevedon and Tickenham (9) and in Leigh Woods (10). It also occurs on Steep Holm. Very rare inland and recorded recently only at Brympton (4), 1974 (CAH), Charlton Adam (5), 1976 (RGR), Cheddar (9) and Writhlington (10), 1960 (PFH).

†**Orobanche maritima** Pugsl. Carrot Broomrape

O. amethystea auct.

Recorded on Brean Down (9) in 1879 (W. B. Waterfall det. A. Bennett) and by later observers, the most recent being I. W. Evans in 1957. Many authorities regard this as only a variety of *O. minor*.

LENTIBULARIACEAE

PINGUICULA L.

Pinguicula lusitanica L. Pale Butterwort

In boggy places. Rare and local. In a few places on Exmoor (1 & 2), near Milverton (3) and, more plentifully than elsewhere in the county, on the Blackdown Hills (3, 4 & 6). Formerly more frequent but now very rare on the peat moors (8).

Pinguicula vulgaris L. Common Butterwort

On boggy and peaty heaths. Believed to be extinct in Somerset until in 1973 it was found in one of the Quantock combes (2) (JGK). Its appearance in an area where it had never been recorded before and near a well-used path must give rise to the suspicion that it was deliberately introduced; it remains to be seen whether it will survive there. Recorded on the peat moors (8) by Sole (1791) it was still said to be abundant in 1886 (D. Fry) after which time it decreased until by 1926 only three or four plants could be found. It has not been reported there since 1928 (WDM). It also grew once near Gurney Slade (10) but the place has long since been drained and the plant lost.

Pinguicula grandiflora Lam. Large-flowered Butterwort

In wet bogs. Introduced in Somerset. In 1969 a small colony was found in a valley bog near Holford (2) (HWB) and in 1970 a larger one in a boggy flush near Oare (2) (CAH). Both colonies are close to paths and there is little doubt that both were deliberate introductions from south-west Ireland where the species is native and plentiful.

UTRICULARIA L.

Utricularia vulgaris L. Greater Bladderwort

In marshland ditches and pools. Quite frequent but local on the Levels in districts 3, 5 and 8. Rare in district 9 and only recorded recently on Cheddar, Kenn and Nailsea Moors. Formerly it occurred in the canals near Bath (10) but has not been seen there this century.

Utricularia neglecta Lehm.

U. major auct.

In similar places to the last. Very rare and only on the peat moors (8) where it was first found in 1902 (D. Fry & JWW). Former flowerless records from the Gordano valley and near Priddy (9) are doubtful as this can only be distinguished with certainty from *O. vulgaris* when in flower.

(**Utricularia intermedia** Hayne has been erroneously included in some Somerset lists.)

Utricularia minor L. Lesser Bladderwort

In marshland ditches and pools. Very rare and only on the peat moors (8) where it was first recorded by Hudson (1778). The records of non-flowering plants in district 9 in White (1912) are considered doubtful. A record by A. J. Yarranton from Stawell (5) has never been confirmed.

VERBENACEAE

VERBENA L.

Verbena officinalis L. Vervain

On roadsides and in dry waste places. Frequent except in the west. Recorded in all districts except 1 and not recently in 6.

LABIATAE

Mentha L.

Mentha pulegium L. Pennyroyal

Edges of ponds and wet commons. Extinct as a native in Somerset but occasionally recorded as a garden escape or casual. It was probably native at Radlet Common, Spaxton (3), where it was collected by Gapper in 1826 but the place has long since been drained. At Walton-in-Gordano (9), 1949 (IWE), it has persisted but probably originated as an escape, as did other occurrences near Wrington (9) 1916 (Miss Todd) and at Bedminster (10).

Mentha arvensis L. Corn Mint

In cultivated ground, open woodland and waste places. Common and recorded in all districts.

Mentha × gentilis L. Bushy Mint

M. arvensis × spicata; M. gracilis Sole; *M. cardiaca* (Gray) Bak.

Formerly grown in cottage gardens and sometimes escaping but very seldom seen recently.

Mentha aquatica L. Water Mint

M. hirsuta Huds.

By streams and in other wet places. Very common throughout the county.

Mentha × verticillata L. Whorled Mint

M. aquatica × arvensis; M. sativa L.

Woodland rides, marshy places and streamsides. Frequent and widely distributed. Not recorded recently in district 5.

Mentha × smithiana R. A. Grah. Tall Mint

M. aquatica × arvensis × spicata; M. rubra Sm. non Mill.

By streamsides and on ditchbanks. Rare. The only confirmed recent records are from Pilton (8), 1970 (CAH), Litton (10), 1979 (RSC), Clandown (10), 1964 (R & IR), and Priston (10), 1963 (R & IR). There are many older records in other districts except 5, 6 and 7, but it does not seem to have persisted in any of these places.

LABIATAE

Mentha × piperita L. Peppermint

M. aquatica × spicata; M. pubescens auct.; *M. citrata* Ehrh.

An escape from cultivation widely naturalized by streams and in waste places. Recorded in all districts but not recently in 6. A hairy variety, which has been misidentified as *M. × dumetorum* Schultes, sometimes occurs; recent records are near Charterhouse (9) and Compton Dando (10). The cultivar nm. **citrata** (Ehrh.) Boivin, which has a scent resembling Eau de Cologne, was found by a roadside ditch on Mendip (9) in 1905 (JWW). In 1925 White said it had lost much of its original odour; the plant growing there now is apparently ordinary Peppermint but whether it belongs to the original clone or is a later arrival is uncertain.

Mentha spicata L. Spear Mint

An alien plant, much cultivated and frequently found in waste places in all districts as a garden escape. A hairy variant, which has been mistakenly identified as *M. longifolia* (L.) Huds., is not uncommon.

× **longifolia** (*M. × villoso-nervata* auct.). A cultivated hybrid which is sometimes found as a garden throw-out, as at West Horrington (8), 1967 (IFG det. R. M. Harley).

Mentha suaveolens Ehrh. Round-leaved Mint

M. rotundifolia auct.

Field borders and waste places. Rare. Perhaps native near Bossington (2). An escape elsewhere. Recently recorded only at Stoke St Gregory (3), 1977 (RSC), Shepton Beauchamp (4), 1973 (CJC det. R. M. Harley) and Bridgwater (5), 1967 (HWB). Often confused with *M. × villosa.*

Mentha × villosa Huds. Large Apple-mint

M. suaveolens × spicata; M. cordifolia auct.; *M. niliaca* auct.

On roadsides and in waste places. A very variable escape from cultivation. Glabrous variants have been known as *M. cordifolia* and are uncommon. The most widespread and frequent form of the hybrid is nm. **alopecuroides** (Hull) Briq. (*M. niliaca* auct.) which has been recorded in all districts except 1 and 3.

LYCOPUS L.

Lycopus europaeus L. Gipsywort

Banks of rivers and ditches, by pools and in other wet places. Common in the lowlands and quite frequent elsewhere except on the higher ground. Not recorded in district 6.

ORIGANUM L.

Origanum vulgare L. Marjoram

Grassland, scrub and roadsides. Common on calcareous soils. Rare elsewhere but recorded in all districts.

THYMUS L.

Thymus pulegioides L. Large Thyme

T. chamaedrys Fr.

In calcareous downland pastures. Frequent in the eastern half of the county especially on the Oolite and Lias. Absent from the west and the lowlands. Not recorded in districts 1 and 2.

Thymus praecox Opiz ssp. **arcticus** (E. Durand) Jalas Wild Thyme

T. drucei Ronn.; *T. serpyllum* auct.

Dry grassland, particularly on hilly calcareous soils. Rather rare in the west except on the coastal Lias. Frequent in the south and very common on the Carboniferous and other limestones of north Somerset.

SATUREJA L.

Satureja montana L. Winter Savory

Once cultivated as a pot-herb. Long naturalized on old walls at Mells Manor (10).

CALAMINTHA Mill.

Calamintha sylvatica Bromf. ssp. **ascendens** (Jord.) P. W. Ball Common Calamint

C. ascendens Jord.; *C. menthaefolia* auct.; *Clinopodium calamintha* auct.

Hedgebanks, roadsides and dry waste places. Common in districts 2, 3 and 9. Rather rare elsewhere but recorded in all districts.

LABIATAE

ACINOS Mill.

Acinos arvensis (Lam.) Dandy — Basil Thyme

Clinopodium acinos (L.) Kuntze; *Calamintha arvensis* Lam.

In stony fields and on limestone rock and rubble. Rare. Only recorded recently at Charlton Mackrell (5), where it has a long history, and on the limestones of districts 9 and 10. Always rare in South Somerset it has not been recorded there since 1941.

CLINOPODIUM L.

Clinopodium vulgare L. — Wild Basil

Calamintha clinopodium Benth.

In rough grassland, on hedgebanks and roadsides. Very common especially on calcareous soils except in the extreme west and in parts of the Levels.

MELISSA L.

Melissa officinalis L. — Balm

By roadsides and in waste places. A frequent escape from cultivation which is sometimes very persistent but is usually found near houses or the sites of demolished cottages. Recorded from all districts except 1 and 6.

SALVIA L.

Salvia verticillata L. — Whorled Clary

A weed of waste places. An alien, established at Lower Marsh, Dunster (2) since 1918 (J. A. Fort). Recorded as a casual in several places in districts 8, 9 and 10 but only seen recently at Tickenham (9), 1967 (ESS).

Salvia verbenaca L. — Wild Clary

S. horminoides Pourr.

Dry banks and grassland on calcareous soils. Frequent in districts 2, 5, 8, 9 and 10. Rare elsewhere. The only recent records in other districts are at Cannington (3), Curry Rivel (4) and Cricket St Thomas (6).

Salvia reflexa Hornem.

A rare casual recorded from a tip at Bedminster (10), 1978 (ALG det. EJC), and a building site in Bath (10), 1979 (RDR).

Salvia viridis L.

S. horminum L.

Garden escape.

MELITTIS L.

Melittis melissophyllum L. Bastard Balm

Hedgebanks and wood borders. Very rare and now only in two localities in the west of district 2. Formerly in a third, near Roadwater, but not recorded there since 1916.

PRUNELLA L.

Prunella vulgaris L. Self-heal

In grassland and waste places. Very common throughout the county.

Prunella laciniata (L.) L. Cut-leaved Self-heal

Grassland on calcareous soil. Very rare and at present known only at Charlton Mackrell (5), 1966 (CAH), and Cheddar (9), though it was formerly recorded at Bruton (8) also. When found at Cheddar by White in 1906 this was thought to be the first British record, but in 1915 C. E. Moss discovered that a specimen in the British Museum herbarium, reported by H. Weaver in 1887 as *P. vulgaris* var. *alba,* was in fact this species. It occurs in several counties in southern Britain and though originally thought to be a native it is now considered to be an introduction from the European mainland.

× **vulgaris** (*P.* × *hybrida* Knaf; *P.* × *intermedia* Link). In calcareous grassland. Variable in colour and in division of the leaves. At both the known localities for *P. laciniata* the hybrid is the more plentiful. At Compton Pauncefoot (5) the hybrid only was found in 1966 by C. A. Howe. Its persistence after *P. laciniata* has disappeared has been reported elsewhere in Britain.

STACHYS L.

Stachys officinalis (L.) Trev. Betony

S. betonica Benth.; *Betonica officinalis* L.

In grassland, woods and on heaths, particularly on neutral or acid soils. Common over most of the county but absent from much of the Levels.

LABIATAE

Stachys arvensis (L.) L. Field Woundwort

A weed in cultivated land. Frequent in the west and formerly elsewhere but now rare in the rest of the county. Not recorded in district 5 nor recently in 7.

[**Stachys germanica** L. was recorded near Portishead (9) (Duck, 1852) and Bristol (Stephens, 1835) but White suggested these were errors for the garden plant **S. byzantina** C. Koch (*S. lanata* Jacq.)]

Stachys palustris L. Marsh Woundwort

By streams and ditches and a weed in cultivated land. Common throughout the county.

× **sylvatica** (*S.* × *ambigua* Sm.). In ditches and cultivated land, often in the absence of the parents. Quite frequent but not recorded recently in the western half of the county.

Stachys sylvatica L. Hedge Woundwort

Hedgebanks, woods and waste places. Very common in all districts.

BALLOTA L.

Ballota nigra L. Black Horehound

Roadsides, hedges and waste ground usually near houses. Common on the lower ground. Not recorded recently in district 1.

LAMIASTRUM Heist. ex Fabr.

Lamiastrum galeobdolon (L.) Ehrend. & Polatsch. Yellow Archangel

Galeobdolon luteum Huds.; *Lamium galeobdolon* (L.) L.

Woods and shady hedgebanks. Very common throughout the county except on the Levels.

LAMIUM L.

Lamium amplexicaule L. Henbit Deadnettle

Cultivated ground and roadsides. Not common but recorded in all districts except 1 and 7.

Lamium hybridum Vill. Cut-leaved Deadnettle

In cultivated ground and waste places. Very rare. Only recorded recently from sandy fields in district 4, on Brean Down (9), 1968 (RMB), and Steep Holm, 1978 (Miss E. Young), though formerly more frequent in district 9. Forms of *L. purpureum* are sometimes reported in error as this species.

Lamium purpureum L. Red Deadnettle

Cultivated ground and waste places. Very common throughout the county.

Lamium album L. White Deadnettle

Hedgebanks, roadsides and waste places. Very common throughout the county except in the wilder parts of Exmoor.

Lamium maculatum L. Spotted Deadnettle

A frequent garden escape which seldom persists for long in the wild.

LEONURUS L.

†**Leonurus cardiaca** L. Motherwort

A garden escape last recorded in the county at Bossington (2) in 1930 having been known there since 1913.

PHLOMIS L.

†**Phlomis fruticosa** L. Jerusalem Sage

A garden plant formerly established on the borders of a plantation on Ball Hill, Stawell (5) where it survived for a good many years but has long since disappeared.

Phlomis russelliana (Sims) Benth. emend Vierhapper Syrian Sage

P. viscosa Poir.

A garden escape well established on a roadside at Mells (10) since 1957 (Mrs. N. Wycherley).

GALEOPSIS L.

Galeopsis angustifolia Ehrh. ex Hoffm. Red Hemp-nettle

G. ladanum auct.

A weed of cornfields on calcareous soils, formerly quite widespread and recorded in all districts except 1 and 6, but which has become very rare recently. Only once recorded since 1960, at Corfe (3), 1978 (Miss K. Wilson).

Galeopsis tetrahit L. sensu lato — Common Hemp-nettle

A weed of arable land and roadsides. Common in all districts. The segregate, **G. bifida** Boenn., appears to be frequent particularly on the peat moors (8).

Galeopsis speciosa Mill. — Large-flowered Hemp-nettle

A weed of cultivated ground which is only a rare casual in Somerset. The only recent records are at Merriott (4), 1974 (SH), and Gare Hill (10), 1969 (PT). There are several records on the peat moors (8) between 1920 and 1954.

NEPETA L.

Nepeta cataria L. — Cat-mint

Hedgebanks and waste places on calcareous soils. Very rare and only recorded recently at Charlton Mackrell (5), 1965 (CAH), Weston-in-Gordano (9), South Stoke and Kelston (10), 1979 (DG), although formerly more widespread.

Nepeta mussinii Spreng. ex Henck.

This and its hybrid (with *N. nepetella* L.?) **N× faassenii** Bergmans ex Stearn are common garden plants and sometimes escape.

GLECHOMA L.

Glechoma hederacea L. — Ground Ivy

Nepeta glechoma Benth.

Hedgebanks, woods and cultivated ground. Very common throughout the county.

MARRUBIUM L.

Marrubium vulgare L. — White Horehound

On Carboniferous Limestone and in sandy places near the coast. Rare and local. On the limestone at Cannington (3), on Mendip from Brean

Down as far inland as Cross (9), on Worle Hill and near Clevedon (9). Formerly recorded on the coast at intervals between Bossington (2) and Brean (9) it is now confined to the stretch between Minehead and Steart (2).

SCUTELLARIA L.

Scutellaria galericulata L. Skullcap

By rivers, streams and ditches and in marshy places. Common throughout the Levels and frequent in the lowlands in other districts. Very rare in the west, the only recent record being by the Barle near Landacre (1), 1971 (MT).

Scutellaria minor Huds. Lesser Skullcap

In bogs and wet peaty places on acid soil. Frequent on the moors and hills of the west and south. Rare in North Somerset and only recorded recently on Blackdown (9) and Downhead (10) on Mendip and at Yarnfield (10) (Wilts). Once abundant on the peat moors (8) according to Sole (1782) but not recorded there since 1822 (Gapper).

Scutellaria altissima L.

S. columnae auct.

A plant from south-east Europe which has become thoroughly naturalized in a wooded valley near Mells (10) where it was first noticed by H. T. Baker in 1929.

TEUCRIUM L.

Teucrium chamaedrys L. Wall Germander

A garden escape which is very persistent when once established. Recent records with dates of first report are on an old wall at Curry Mallet (3), 1922 (Preb J. Hamlet), on rocks at Shepton Montague (8), 1966 (CAH), and at Sand Point (9), 1967 (R & IR), and on a paved footpath at Odd Down, Bath (10), 1970 (H. Bond).

Teucrium scorodonia L. Wood Sage

Woods, heaths and dry stony places, preferring acid soils. Very common in the west and south and common on Mendip and the other Carboniferous Limestone and sandstones in the north. Rare on the Oolite and Lias and on the Levels.

LABIATAE

AJUGA L.

Ajuga reptans L. Bugle

Woods, hedges and damp grassland. Very common in all districts.

PLANTAGINACEAE

PLANTAGO L.

Plantago major L. Greater Plantain

Very common throughout the county in waste places and on roadsides.

Plantago media L. Hoary Plantain

Calcareous grassland. Very common in the eastern half of the county except on the Levels where it is scarce. In the west it is frequent on the coast from Bossington to Steart (2). Rare in district 3 west of Taunton and not recorded in district 1.

Plantago lanceolata L. Ribwort Plantain

Very common in meadows and other grassland and in waste places throughout the county.

Plantago maritima L. Sea Plantain

Common on mud along the coast from Porlock (2) to Portbury (10) and up the tidal river estuaries.

Plantago coronopus L. Buck's-horn Plantain

On rocks and sand near the sea and in dry places inland on sand and gravel. Common on the coast from Porlock (2) to Portbury (10) and by the Avon as far up as Bristol (10). Also at the north end of Quantock (2) and on Mendip as far inland as Cross (9). Very rare elsewhere inland and only seen recently at Leigh Hill (3 & 6) and Chiselborough (4) though recorded formerly in a few other places in these districts. A perennial form, var. **sabrinae** Baker fil. & Cardew, with broader, fleshy leaves, is common on Steep Holm where it was first collected about 1845 by the Rev T. Butler **(BM)** and was considered by Druce to be a distinct species, *P. sabrinae,* and by F. N. Williams to be conspecific with *P. serraria* L., a Mediterranean plant. It occurs rarely on the rocks of Brean Down also (9).

Plantago arenaria Waldst. & Kit.

P. indica L.; *P. psyllium* L. nom. ambig.

A rare casual which appeared in abundance on the sand dunes at Burnham (8) in 1859 and persisted there for about ten years. It has since been seen in three other places, most recently on a rubbish tip at Yeovil (4), 1961 (VIR det. CCT).

LITTORELLA Berg.

Littorella uniflora (L.) Aschers. Shoreweed

L. juncea Berg.

On the margins of ponds and reservoirs. Very rare. It occurs by the Barle in a few spots between Cow Castle and Withypool (1), an unusual habitat in which it was first recorded by Marshall in 1918. The only other localities in which it has been seen recently are by ponds at Holford (2) and on Blackdown (9), and by the reservoirs at Clatworthy (3), Durleigh (3) and Blagdon (9). Formerly recorded also at Chard Reservoir (4), by a pool near Burnham (8) and by Ray (1670) on the peat moors (8) but not seen there since.

CAMPANULACEAE

WAHLENBERGIA Schrad.

Wahlenbergia hederacea (L.) Reichb. Ivy-leaved Bellflower

Cervicina hederacea (L.) Druce

Wet moors and damp woodland rides on acid soils. Common on Exmoor and the Brendons (1 & 2) and frequent at the northern end of Quantock (2). Elsewhere it occurs rarely on the peat moors (8) and on the Greensand of the eastern border (7 & 8) where it is apparently decreasing. Formerly it extended as far north as Gare Hill (10) but has not been seen there recently. The absence of this species from the Blackdown Hills is remarkable.

CAMPANULA L.

Campanula latifolia L. Giant Bellflower

Only a garden escape in Somerset but well naturalized in hedges and by rivers and having the appearance of a native in a few places, as near Dulverton (1), Cutcombe (2) and Bruton (8).

CAMPANULACEAE

Campanula trachelium L. Nettle-leaved Bellflower

In hedges, scrub and open woodland on calcareous soils. Common on the Carboniferous Limestone of Mendip and elsewhere in the north of the county and on the Oolite from Bratton Seymour (8) to Bath (10). Occasional on the chalk of the Wiltshire border (7). Doubtfully native in South Somerset where the only recent record is in an old quarry near Minehead (2) where it has been known since 1914 (A. H. Wolley-Dod). It has gone from another station near Roadwater (2) where it was recorded at the same time.

Campanula rapunculoides L. Creeping Bellflower

A rare garden escape in Somerset. It has become established on the disused railway line near Shapwick (8) where it was first noticed in 1942 (C & NS).

Campanula persicifolia L. Peach-leaved Bellflower

Garden escape.

Campanula glomerata L. Clustered Bellflower

Calcareous grassland. It is near the western limit of its English distribution in Somerset. Fairly frequent on the Oolite round Bath (10) and occasional on the Jurassic formations as far south as Shepton Montague (8 & 10); frequent also on the eastern chalk (7) (Wilts). Elsewhere in North Somerset it occurs only on the Lias near Butleigh (5) and on the Carboniferous Limestone of Mendip near Shipham (9). Very rare in South Somerset and recorded recently only at South Cadbury (5), 1965 (CAH), though there is an old record from the Lias near Quantoxhead (2) by H. S. Thompson in Murray (1896).

Campanula rotundifolia L. Harebell

On banks and in dry calcareous grassland. Common on the chalk (7) (Wilts), the Carboniferous Limestone in the north of the county (8, 9 & 10) and on the Bath Oolite (10). Very rare elsewhere, the only recent records being at Ham Hill (4), near Charlton Horethorne (5) and near Puriton (8), 1972 (TF). Old records for Grabbist Hill (2) and Cothelstone Hill (3) have not been confirmed since.

Campanula patula L. Spreading Bellflower

Shady banks and open woodland. Very rare and now only near East

Harptree (10) where it was first reported by Rutter (1829). Casual only and impermanent in a number of other localities except perhaps Creech Hill, Bruton (8) where it survived for thirty years but has not been seen since 1937.

Campanula alliariifolia Willd. Cornish Bellflower

A garden escape well established with other aliens on railway banks near Langport (5), 1972 (CAH).

Campanula medium L. Canterbury Bells

Garden escape.

Campanula portenschlagiana Schultes

Commonly grown in gardens and occasionally established on walls and in path crevices near houses.

LEGOUSIA Durande

Legousia hybrida (L.) Delarb. Venus's Looking-glass

Specularia hybrida (L.) A.DC.

In cornfields on calcareous soils. Rare and decreasing, though it is still frequent on the Lias near Somerton (5) and on the Oolite south of Bath (10). Elsewhere it has only been seen recently near Watchet (2), Ham Hill (4), 1974 (SH), and Brewham (8), 1965 (CAH), though formerly recorded more widely and in districts 3, 7 and 9 also.

JASIONE L.

Jasione montana L. Sheep's-bit

In dry grassland and on rocky banks on acid soils. Common on Exmoor and the Brendons and on the Blackdown Hills; less so on the Quantocks. Reported near Clevedon (9) in 1896 (Mrs. Lainson) and confirmed in 1912 by D. Williams who sent a specimen collected in 1908 to White; the spot had been built over by 1918 according to Miss Livett. Apart from this one there are no other reliable records from North Somerset.

(**Lobelia dortmanna** L. was planted in ponds at Culmhead (6) about 1918 but has not survived.)

RUBIACEAE

SHERARDIA L.

Sherardia arvensis L. Field Madder

A weed in cultivated and waste land and in grassland on calcareous soils. Common and widespread but not recorded recently in district 1.

PHUOPSIS (Griseb.) Hook.f.

Phuopsis stylosa (Trin.) B. D. Jackson

Garden escape.

ASPERULA L.

Asperula cynanchica L. Squinancywort

Calcareous downs. Locally common on the Lias from Charlton Mackrell (5) to Walton (8), on the eastern chalk (7) (Wilts), on Mendip from Bleadon (9) to Ebbor (8), and on Worle Hill (9). Occasional on the Oolite from Creech Hill, Bruton (8) to Bannerdown (10). Formerly recorded also near Clevedon (9) but not seen there for many years. Absent from South Somerset; an occurrence at Galmington, near Taunton (3), 1918 (WW), must have been a casual introduction.

Asperula arvensis L.

An uncommon casual.

CRUCIATA Mill.

Cruciata laevipes Opiz Crosswort

C. chersonensis auct.; *Galium cruciata* (L.) Scop.

Hedgebanks, grassy roadsides and by streams. Rather scarce in the west and south but locally common in North Somerset. Not recorded recently in district 5.

GALIUM L.

Galium odoratum (L.) Scop. Woodruff

Asperula odorata L.

In woods and shady lanes. Common over much of the county but scarce on the Levels where there are few suitable habitats.

Galium mollugo L. Hedge Bedstraw

inc. *G. album* Mill.

On hedgebanks, roadsides and in rough grassland. Very common everywhere except on Exmoor. An early-flowering form with narrower leaves and more-ascending branches is not uncommon on the Lias, Oolite and in other calcareous grassland and has been separated as ssp. **erectum** Syme (*G. erectum* Huds.) but is rather variable.

× **verum** (*G.* × *pomeranicum* Retz.; *G.* × *ochroleucum* Wolf ex Schweigg. & Koerte) occurs occasionally on roadsides and in calcareous grassland. It has been recorded more often on the Lias of district 5 and on Mendip (9) than elsewhere.

Galium verum L. Lady's Bedstraw

In dry grassland, waste places and on sand dunes. Very common over much of the county but rather scarce in the west. Prefers calcareous soils.

Galium saxatile L. Heath Bedstraw

G. harcynicum Weigel; *G. hercynicum* auct.

On dry banks and heaths on acid soils. Very common on Exmoor and the Brendon, Blackdown and Quantock Hills. Also on the eastern Greensand, the peat moors (8) and on Mendip. Away from these areas it occurs only in a few spots where leaching has produced locally acid conditions, for example, on the top of Solsbury, near Bath (10).

Galium fleurotii Jord.

G. pumilum auct.; *G. sylvestre* auct.; *G umbellatum* auct.; *G. asperum* auct.

Calcareous screes and cliffs. Confined in Britain to a small area of Mendip (9) mainly round Cheddar but stretching from Shipham to Ebbor. Since it was first noticed by Borrer this plant has been called *G. pumilum* or one of its synonyms but in 1951 Sterner considered it was distinct from this more widely distributed plant and suggested it was a separate variety or subspecies. Dr K. M. Goodway, who was working on the group at that time, was reluctant to recognize it as a distinct taxon but a note about the Cheddar population was included in Clapham's account in the second edition of the *Flora of the British Isles* (1962). Later Dr F. Ehrendorfer, who wrote the account of *Galium* for *Flora Europaea 4* (1976), considered it was identical with *G. fleurotii* which had been described from

French material by Jordan in 1849. A specimen dated 1853 from Steep Holm (collector unknown) is in the British Museum herbarium, but this must have been a casual introduction to the island which did not persist.

Galium palustre L. Marsh Bedstraw

In marshes, ditches, by rivers and ponds and in other wet places. Very common throughout the county. **G. elongatum** C. Presl (*G. palustre* ssp. *elongatum* (C. Presl) Lange) is now considered a separate species; both appear common in Somerset but the distributions of the two taxa have not been worked out.

Galium uliginosum L. Fen Bedstraw

In marshy and peaty places. Not common and appears to be decreasing especially in the north of the county. It occurs in all districts but the records are thinly scattered except on the peat moors (8) where it is still frequent.

† **Galium tricornutum** Dandy Corn Cleavers

G. tricorne Stokes pro parte

A cornfield weed formerly widespread and recorded in all districts except 1 and 6. It had become very rare by 1930 and was last seen in the county in 1955 at Compton Dundon (5) (EJH).

Galium aparine L. Goosegrass; Cleavers

In hedges and a weed in cultivated ground. Very common throughout the county.

† **Galium spurium** L. False Cleavers

G. vaillantii DC.

A weed of cultivated ground and waste places usually only a casual in Britain but established on the peat moors (8) from 1913, when it was first noticed by Marshall, until 1946, since when it has not been seen.

† **Galium parisiense** L. Wall Bedstraw

G. anglicum Huds.

Recorded at Charlcombe (10) by Babington (1834) and "on old walls near

Bath" by T. B. Flower in 1883, but both Murray and White suggested these records were errors.

RUBIA L.

Rubia peregrina L. Wild Madder

In hedges, wood borders and scrub in rocky places. Common near the coast between Bossington (2) and Bridgwater (3), at Brean Down and Weston-super-Mare (9), and between Clevedon (9) and Bristol (10). Usually a sub-maritime plant in Britain but in Somerset it occurs up to ten miles inland being common on the Lias of districts 3, 4, 5 and 8 between West Buckland (3) and Barton St David (8). It also occurs on the Carboniferous Limestone of Mendip and Broadfield Down (9). An isolated record at Withypool (1), 1971 (JGK), was probably of an introduction.

CAPRIFOLIACEAE

SAMBUCUS L.

Sambucus ebulus L. Danewort; Dwarf Elder

Roadsides and hedges, usually not far from villages. Rare and usually only in small patches. Known in about a dozen places scattered about the county. Believed to be a relic of ancient cultivation but long established in some of its localities, for example, at Charlton Mackrell (5) where it was collected by Gapper in 1822.

Sambucus nigra L. Elder

Hedges, woods and waste places. Very common throughout the county.

VIBURNUM L.

Viburnum lantana L. Wayfaring Tree

Hedges and woods. Very common on calcareous soils so very rare in the west except on the Lias of district 2. Not recorded except as an introduction in district 1.

Viburnum tinus L. Laurustinus

Garden escape.

CAPRIFOLIACEAE

Viburnum opulus L. Guelder Rose

Woods and hedges. Rather common throughout the county.

SYMPHORICARPOS Duham.

Symphoricarpos albus (L.) S. F. Blake Snowberry

S. rivularis Suksd.; *S. racemosus* Michx

Frequently planted in hedges and often naturalized in copses and by streams.

LONICERA L.

Lonicera xylosteum L. Fly Honeysuckle

An uncommon garden escape.

Lonicera periclymenum L. Honeysuckle

Hedges and woods. Very common throughout the county.

Lonicera caprifolium L. Perfoliate Honeysuckle

Garden escape occasionally naturalized in woods or hedges. Long established in Brassknocker Wood near Bath (10) but not seen there recently.

Lonicera nitida Wils.

A common hedging plant often found as a throw-out from gardens.

LEYCESTERIA Wall.

Leycesteria formosa Wall. Himalayan Honeysuckle

Planted in gardens, and in woodland as food for pheasants, and quite often found as a bird-sown escape.

ADOXACEAE

ADOXA L.

Adoxa moschatellina L. Moschatel

On shady hedgebanks and in woods. Common in many parts of the county and recorded in all districts but rare in central Somerset and absent from the Levels.

VALERIANACEAE

VALERIANELLA Mill.

Valerianella locusta (L.) Laterr. Common Cornsalad; Lamb's Lettuce

V. olitoria (L.) Poll.

A weed of cultivated and waste ground on light or sandy soils. Recorded in all districts except 1. Still quite frequent but apparently decreasing.

Valerianella carinata Lois. Keeled-fruited Cornsalad

Railway banks, quarries, walls and a weed in gardens. First recorded in Somerset at the old quarries on Hampton Down (10) in 1887 by the Rev W. O. Wait it remained very rare (or perhaps overlooked) until quite recently, but since 1965 it has been recorded in more than forty tetrads covering all districts except 1 and 6 for which there are older records.

Valerianella rimosa Bast. Broad-fruited Cornsalad

V. auricula DC.

A cornfield weed; always rare in Somerset it has only been recorded in two places since 1931, at Nether Stowey (3), 1946 (WW), and at Charlton Mackrell (5) where it has been known since 1938 (FKM) and still survives in very small quantity. Formerly recorded in districts 4, 8 and 10 also.

Valerianella eriocarpa Desv. Hairy-fruited Cornsalad

A rare introduced annual weed of dry open habitats which was recorded in one spot on slopes above Cheddar (9) in 1935(C & NS & H. W. Pugsley) but did not persist. It was found for the second time in the county at Charlton Mackrell (5) in 1979 (R & IR) on waste ground bordering a quarry.

Valerianella dentata (L.) Poll. Narrow-fruited Cornsalad

A cornfield weed formerly quite common but decreasing and now rare. Only seen recently at Horner (2), 1974 (CJG), Charlton Mackrell (5), 1977 (R & IR), Pitcombe (8), 1962 (CAH), and in several spots round Bath (10). Formerly recorded in districts 3, 4, 7 and 9 also.

VALERIANACEAE

Valeriana L.

Valeriana officinalis L. Common Valerian

inc. *V. sambucifolia* auct.

Woods and grassy places. Common on Exmoor, the Blackdown Hills, the peat moors (8) and Mendips, and in the north and east of the county generally. Rather rare elsewhere including the Quantocks and the Levels but recorded in all districts. Rather variable. Two genotypes occur but they cannot be distinguished morphologically.

Valeriana pyrenaica L. Pyrenean Valerian

First recorded in a plantation above Dulverton (1) in 1883 by Murray, it had spread quite extensively in the Barle valley by 1891 and was also reported from the Haddeo valley above Bury (1) (Rev H. M. Trott). It has become thoroughly naturalized in these places but has not spread further.

Valeriana dioica L. Marsh Valerian

In marshy woods and meadows. Much less common than in the past due to the drainage of many localities, it is now rather rare except on the Blackdown Hills (3, 4 & 6) and on the eastern border between Bruton and Frome (7, 8 & 10). Recorded in all districts but not recently in 2, it is absent from the coastal lowlands and much of the Levels except on the peat moors (8).

Centranthus DC.

Centranthus ruber (L.) DC. Red Valerian

Kentranthus ruber auct.

Originally introduced but long established in the wild on cliffs, in quarries, on banks and old walls as well as in towns and villages where it is more obviously an escape. Common in all districts.

DIPSACACEAE

Dipsacus L.

Dipsacus fullonum L. Wild Teasel

D. sylvestris Huds.

Rough grassland, roadsides, ditchbanks and waste places. Common throughout the county except on the higher ground of Exmoor, the

Brendons, Blackdowns, Quantocks and Mendip. Not recorded recently in district 1.

Dipsacus sativus (L.) Honck. Fuller's Teasel

D. fullonum sensu Mill. non L.

Still grown on a small scale round Fivehead and Curry Rivel (4) and strays from cultivation are very occasionally met with on grassy roadsides in the district but they do not persist. Formerly grown in the north of the county as well but not seen there except as a rare casual on rubbish tips for over a century.

Dipsacus pilosa L. Small Teasel

Damp shady places by streams and in woods. Still frequent in the east of the county in districts 8 and 10 but now very rare elsewhere. Formerly more common but there are few recent records in districts 2, 3, 7 and 9 and none in 4 or 5. Not recorded in district 1.

KNAUTIA L.

Knautia arvensis (L.) Coult. Field Scabious

Scabiosa arvensis L.

Rough grassland, roadsides, arable land and waste places. Common except in the extreme west on Exmoor.

SCABIOSA L.

Scabiosa columbaria L. Small Scabious

Dry grassland and stony banks on calcareous soils. Frequent on the Lias in districts 2, 3 and 4; common on the Polden Hills (5 & 8), the chalk on the Wiltshire border (7), the Carboniferous Limestone of Mendip and other outcrops in the north (8, 9 & 10), and the Oolite round Bath and Frome (10). Rare elsewhere and absent from district 1.

SUCCISA Haller

Succisa pratensis Moench Devil's-bit Scabious

Scabiosa succisa L.

Marshy pastures, moors and open woods, preferring acid soils but not confined to them. Widespread and common except on the Levels where it only occurs on the peat moors (8) and in a few other places where the peat is near the surface.

COMPOSITAE

Guizotia Cass.

Guizotia abyssinica (L.f.) Cass.

A bird-seed alien increasingly found on rubbish tips.

Bidens L.

Bidens cernua L. Nodding Bur-marigold

In ditches, by reservoirs and ponds. Common on the peat moors (8). Rare elsewhere; there are a few scattered records in districts 3, 4, 5, 7 and 9. Absent from the west of the county (1, 2 & 6).

Bidens tripartita L. Trifid Bur-marigold

In similar places to the last species but also by rivers. More frequent and widespread. Recorded in all districts except 1.

Galinsoga Ruiz & Pav.

Galinsoga parviflora Cav. Gallant Soldier

A rare casual, first recorded in the county in 1951 (W. E. Palmer) on a bombed site in Yeovil (4) but it did not persist there.

Galinsoga ciliata (Raf.) Blake Shaggy Soldier

An introduced weed of nurseries, gardens and town roadsides, first recorded in Somerset in 1944 in nursery gardens at Bath (10) (E. H. Lubbock). From there it spread and can still be found in a few places about the city. It has since been recorded in a few other localities in districts 3, 4, 9 and elsewhere in 10, but it is still very uncommon.

Helianthus L.

Helianthus annuus L. Sunflower

A garden escape, frequent on rubbish tips.

Helianthus rigidus (Cass.) Desf.

Garden escape.

TAGETES L.

Tagetes minuta L.

A rare casual recorded at Yeovil (4), ca. 1959 (VIR) and at Hinton Charterhouse, (10), 1963 (C. D. Brickell).

AMBROSIA L.

Ambrosia artemisiifolia L. Ragweed

Introduced with imported grain and first recorded in Somerset in 1900 at Portishead Dock (9) (JWW). Still an occasional casual where fowls have been fed.

XANTHIUM L.

Xanthium strumarium L. Cocklebur

A rare casual in Somerset though commoner elsewhere in the country, not reported until 1918 at Brislington (10) (Miss M. Cobbe). Ssp. **italicum** (Moretti) D. Löve was found at Bedminster (10), 1928 (CIS). A specimen found on a tip at Glastonbury (8) in 1970 by C. A. Howe was identified as the segregate **X. saccharatum** Wallr. by Professor F. J. Widder of Graz.

Xanthium spinosum L. Spiny Cocklebur

A rare casual of rubbish tips, etc., first reported in the county in 1905 at Portishead (9) (JWW). The only recent record was on a tip at Glastonbury (8), 1970 (CAH & CES).

SENECIO L.

Senecio jacobaea L. Common Ragwort

Pastures, roadsides and waste places. Very common in all districts.

Senecio aquaticus Hill Marsh Ragwort

By streams and in marshy fields. Frequent in all districts particularly in the lowlands and on acid soils.

× **jacobaea** (*S.* × *ostenfeldii* Druce) has occurred in the county notably in the Gordano valley (9).

COMPOSITAE

Senecio erucifolius L. Hoary Ragwort

Grassy roadsides and rough grassland. Common on calcareous and clay soils. Rare in the west except near the coast and not recorded in district 1.

Senecio squalidus L. Oxford Ragwort

Waste places, walls and railway ground. Introduced on walls in Taunton (3) about 1820 by the Rev W. Tuckwell, it was not recorded elsewhere in the county until 1904 when it was found at Portishead (9) (W. Hosking) **(BM).** For the next 25 years new records were confined to the area between Bristol (10) and Weston-super-Mare (9), but in 1929 it was collected at Frome **(BM).** In 1944 it was at Bath (10), increasing (DEC), and in 1956 at Watchet (2), Yeovil (4) and near Wells (8). It is now fairly widespread and recorded in all districts except 1. Most records are near railways but it is now found away from them too. It is most plentiful around Minehead (2), Taunton (3), Portishead (9), Bristol and Bath (10).

× **vulgaris** (*S.* × *baxteri* Druce). Although the parent species frequently grow together on waste ground hybrids have seldom been recorded in the county but have been seen, as at Keynsham (10) in 1973, together with the parents.

Senecio sylvaticus L. Heath Groundsel

On dry heaths and peaty ground on acid soils. Quite common on Exmoor, the Brendons and Quantocks and on dry parts of the peat moors (8). Rare elsewhere but recorded in all districts.

× **viscosus** (*S.* × *viscidulus* Scheele) has been recorded on the peat moors (8) where the parents grow together.

Senecio viscosus L. Sticky Groundsel

A weed of waste places especially by railways. First recorded in Somerset in 1876 at Clevedon (9) (Mrs. Lainson), it was said by White to be increasing rapidly but it has since declined in the north of the county. Recent records are rather thinly scattered about in all districts except 1 and 2 where it occurred formerly but has not been reported since 1950. More frequent in the south-east (4, 5, 7 & 8) and on the peat moors (8) than anywhere else.

Senecio vulgaris L. Groundsel

A weed of cultivated and waste ground, very common everywhere. The rayed variety, f. **radiatus** Hegi, is considered by some authorities to have been derived by introgressive hybridization with *S. squalidus* but White (1912) gives several records before the latter had been reported in Somerset. There are a number of recent records in districts 2, 4, 8, 9 and 10, and some of these are in places where *S. squalidus* has not yet been recorded.

Senecio fluviatilis Wallr. Broad-leaved Ragwort

S. sarracenicus L. pro parte

Originally introduced but very long established by streams and rivers. First recorded by Bobart about 1680 by a small stream between Wells and Glastonbury (8) (Ray, 1724) where it still grows. It is most plentiful in district 8 east of Street by the Brue and its tributary the Alham. It also occurs in district 10 by the Avon and its tributaries the Chew, Cam and Wellow Brooks and the Frome. In a few other places in districts 5 and 9 it is probably a more recent escape.

Senecio integrifolius (L.) Clairv. Field Fleawort

Only on the chalk of White Sheet Hill (7) (Wilts) on the boundary of V.C.6.

Senecio bicolor (Willd.) Tod. ssp. **cineraria** (DC.) Chater Silver Ragwort

S. cineraria DC.

Naturalized on shingle and sea cliffs between the Devon border and Hurlstone Point (2) where it was first reported by N. G. Hadden in 1924. An occasional garden escape elsewhere.

DORONICUM L.

Doronicum pardalianches L. Leopard's-bane

A garden escape which has become thoroughly naturalized in a few places such as near Dulverton (1) and Flax Bourton (9) but is less firmly established in others.

Doronicum plantagineum L. Plantain-leaved Leopard's-bane

Garden escape.

COMPOSITAE

TUSSILAGO L.

Tussilago farfara L. Colt's-foot

Roadsides, railway banks, waste ground and poor fields. Very common especially on clay soils and on recently disturbed ground.

PETASITES Mill.

Petasites hybridus (L.) Gaertn., Mey. & Scherb. Butterbur

P. officinalis Moench

By streams and in damp meadows. Common in much of the county but scarce on the Levels. Recorded in all districts.

Petasites fragrans (Vill.) C. Presl Winter Heliotrope

A garden escape established on roadsides throughout the county most frequently near towns and villages but sometimes far from houses.

CALENDULA L.

Calendula officinalis L. Pot Marigold

A garden escape common on rubbish tips.

INULA L.

Inula helenium L. Elecampane

Originally an escape from cultivation but now quite naturalized on field borders in many parts of the county. Rare in the west. Recorded in all districts but not recently in 3.

Inula conyza DC. Ploughman's Spikenard

Grassy slopes, banks and roadsides, preferring calcareous soils. Common on the Lias and the limestones of North Somerset but rather sparse elsewhere. Not recorded in districts 1 or 6.

Inula crithmoides L. Golden Samphire

Only on the cliffs of Steep Holm where it was first collected by Banks in 1773 **(BM).** A single plant was found in the Berrow salt marsh (9) in 1955 (AJW) but did not long survive there.

DITTRICHIA W. Greuter

Dittrichia viscosa (L.) W. Greuter

Inula viscosa (L.) Ait.

A very rare casual found in 1978 on a building site at Yeovil (4) (CJC det. EJC).

Dittrichia graveolens (L.) W. Greuter

Inula graveolens (L.) Desf.

A wool adventive found on a tip at Glastonbury (8), 1970 (CAH & JGK det. JEL).

PULICARIA Gaertn.

Pulicaria dysenterica (L.) Bernh. Common Fleabane

By ditches, in damp fields and on roadsides. Very common throughout the county except on the higher parts of Exmoor.

FILAGO L.

Filago vulgaris Lam. Common Cudweed

F. germanica (L.) L.

In dry rocky and sandy places. Formerly frequent and recorded in all districts it has now become rather rare and has been seen recently only on Minehead Warren and a few other places in district 2, on Crook Peak and Wavering Down on Mendip (9) and in an old quarry at Downhead (10).

LOGFIA Cass.

†**Logfia minima** (Sm.) Dumort. Small Cudweed

Filago minima (Sm.) Pers.

A plant of dry, sandy, acid soils always very rare in Somerset and confined to a few localities in districts 2, 9 and 10, it was last recorded on the spoil heap of a disused coal mine at Clutton (10) in 1958 (NYS).

OMALOTHECA Cass.

Omalotheca sylvatica (L.) Schultz Bip. & F. W. Schultz Heath Cudweed

Gnaphalium sylvaticum L.

Dry heathy ground on acid soils. Very rare and decreasing. Formerly recorded in all districts except 5, 6 and 8, it has only been seen once recently, at Nailsea (9), 1965 (ESS).

Filaginella Opiz

Filaginella uliginosa (L.) Opiz — Marsh Cudweed

Gnaphalium uliginosum L.

A weed of poorly-drained cultivated ground and roadsides where water has lain. Common on the acid soils of the west and south and quite frequent elsewhere.

Anaphalis DC.

Anaphalis margaritacea (L.) Benth. — Pearly Everlasting

Originally a garden escape first recorded in the wild in Somerset in 1883 but now well established in disused quarries, on old railway tracks and other waste ground in several places on calcareous soils in districts 8, 9 and 10. Old records in districts 1 and 2 did not persist.

Antennaria Gaertn.

†**Antennaria dioica** (L.) Gaertn. — Mountain Everlasting

A northern plant of rocky heaths formerly found in very small quantity in a few places on the Carboniferous Limestone of districts 9 and 10, and last reported at Goblin Combe (9) in 1926 (Miss H. M. Dixon). Once recorded from the peat moors (8), 1913 (C. Perrens). It appears to have become very rare in southern England generally.

Solidago L.

Solidago virgaurea L. — Golden Rod

In rocky places on acid soils. Common in the west on Exmoor, the Brendons and the Quantocks, and on the sandstone of districts 9 and 10. Rare elsewhere. Not recorded in district 5 nor recently in 7.

Solidago canadensis L.

A garden escape first recorded in the county in 1946 at Farrington Gurney (10) (JPMB). Now widely established in waste places, on roadsides and railway banks, although it is still rather rare in the west and not yet recorded in districts 1, 3 and 6.

Solidago gigantea Ait.

An occasional garden escape.

ASTER L.

Aster tripolium L. Sea Aster

On mudbanks by the sea and up the tidal river estuaries from Porlock Weir and Hinkley Point (2) to Bristol (10). Common. The var. **discoideus** Reichb. is equally as common as the rayed variety; both grow together.

Aster novi-belgii L. Michaelmas Daisy

A widespread garden escape commonly naturalized on roadsides and in waste places, often far from houses, in all districts. Although briefly mentioned by White (1912) there are no other published Somerset records so it is not possible to establish when the spread took place. Nearly all the wild plants are the type of this species; more showy garden varieties and hybrids and other cultivated species, including **A. lanceolatus** Willd., are sometimes reported but do not persist for long.

Aster linosyris (L.) Bernh. Goldilocks

Linosyris vulgaris Cass. ex DC.; *Crinitaria linosyris* (L.) Less.

On limestone rocks near the sea. Very rare. Formerly on rocks above Worle and Weston-super-Mare (9) where it was collected by W. Christy in 1830, it was last recorded there in 1857 by Jenyns. It was believed to be extinct in the county until found in 1904 by G. C. Druce some miles away, and in yet another spot in 1965 (PJMN). Although Druce's locality was believed to be a new one, in 1956 Mrs. B. Welch reported that a specimen collected there in 1813 by Dr Wollaston was in the British Museum herbarium.

ERIGERON L.

Erigeron acer L. Blue Fleabane

On dry sandy ground and on walls. Rather common in North Somerset. Less so in the west and south but recorded in all districts except 1 and 6 and not recently in 7.

Erigeron karvinskianus DC. Mexican Fleabane

E. mucronatus DC.

A garden escape widely naturalized on walls. There are no published records for the county before 1953 [Wembdon (3), (EJH)] but it is now recorded in all districts except 1, 6 and 7.

CONYZA Less.

Conyza canadensis (L.) Cronq. Canadian Fleabane

Erigeron canadensis L.

An introduced weed of waste ground first reported in the county in 1884 (D. Williams) and confined to the Bristol area (10) until the 1920s when it was still rare in Somerset but reported as far away as Highbridge (8). It spread rapidly during and after the Second World War, particularly by railways and on the bombed sites of Bristol and Bath, and has now been recorded in all districts except 1 and 6.

× **Erigeron acer.** This very rare hybrid was recorded at Ashton Gate (10) in 1911 (IMR) and persisted there for many years (CIS, 1932). It was seen again at Brislington (10) in 1931 (HST) but has not been recorded anywhere more recently.

Conyza bonariensis (L.) Cronq.

Erigeron bonariensis L.

A rare casual found on a rubbish tip at Minehead (2), 1960-62 (J. I. Robbins det. Kew).

Conyza sumatrensis (Retz) E. Walker

C. floribunda Kunth

Another rare casual, found on a building site at Yeovil (4), 1978 (CJC det. EJC).

BELLIS L.

Bellis perennis L. Daisy

A weed of lawns and grasslands. Very common everywhere.

EUPATORIUM L.

Eupatorium cannabinum L. Hemp Agrimony

By streams and ditches and in damp grassy places. Very common throughout the county except in the higher parts of Exmoor.

ANTHEMIS L.

Anthemis tinctoria L. Yellow Chamomile

Garden escape.

Anthemis cotula L. Stinking Chamomile

A weed of arable land, common on the Lias and Oolite. Rather scarce elsewhere but recorded in all districts.

Anthemis arvensis L. Corn Chamomile

Once a cornfield weed as in other parts of the country it had become a very rare casual by the end of the nineteenth century. There have been very few recent records.

Anthemis punctata Vahl. ssp. **cupaniana** (Tod. ex Nyman) R. Fernandes

A. cupaniana Tod. ex Nyman

Garden escape, well established on railway banks at Curry Rivel (3), first noticed 1971 (CAH) and still there.

CHAMAEMELUM Mill.

Chamaemelum nobile (L.) All. Chamomile

Anthemis nobilis L.

In damp grassy places on sandy and acid soils. Formerly frequent in the west but recorded recently only at Bossington and Holcombe (2). Its disappearance from other places in districts 1 to 4 must be attributed to disturbance of the commons and other marginal land where it used to grow. In North Somerset the only native record was on Brean Down (9) where it was last seen in 1896. It has occurred as a casual or garden throw-out in a few other places, most recently at Bruton (8), 1963 (CAH).

ACHILLEA L.

Achillea millefolium L. Yarrow

Grasslands, roadsides and lawns. Very common everywhere.

Achillea ptarmica L. Sneezewort

Damp meadows and marshes. Common on the Blackdown Hills (3, 4 & 6) and on the eastern borders (7, 8 & 10). Rare elsewhere. Not recorded recently in district 2. The double-flowered form is grown in gardens and sometimes escapes.

Achillea distans Waldst. & Kit. ex Willd. ssp. **tanacetifolia** Janchen

A. tanacetifolia All.

A rare casual from south Europe naturalized on a disused railway track near Highbridge (8), 1968 (JVM det. BM).

MATRICARIA L.

Matricaria maritima L. Seaside Mayweed

Tripleurospermum maritimum (L.) Koch sens. strict.

Common on the shore from Porlock (2) to Portbury (10).

Matricaria perforata Mérac Scentless Mayweed

M. inodora L.; *Tripleurospermum maritimum* ssp. *inodorum* (L.) Hyland. ex Vaarama

A weed of cultivated ground and waste places. Very common in all districts except on the higher parts of Exmoor.

CHAMOMILLA S. F. Gray

Chamomilla recutita (L.) Rauschert Wild Chamomile; Scented Mayweed

Matricaria recutita L.; *M. chamomilla* L. pro parte

A weed of cultivated ground and roadsides which has much increased during the last fifty years and is now very common particularly in the lowlands. Recorded in all districts but still rather scarce on Exmoor.

Chamomilla suaveolens (Pursh) Rydb. Pineapple Weed

Matricaria matricarioides (Less.) Porter pro parte; *M. discoidea DC.; M. suaveolens* (Pursh) Buchen., non L.

Roadsides, field gateways and farmyards. Introduced from North America and now very common everywhere. First recorded in the county at Portishead (9) in 1903 (JWW) it had spread considerably but was still very local by 1914 (ESM); its main spread came later with the increase in motor traffic. Muddy tyres and boots are still the main dispersal agents.

Chamomilla aurea (Loefl.) Gay ex Cosson & Kralik

Matricaria aurea (Loefl.) Schultz Bip.

A casual found on a rubbish tip at Yeovil (4), 1963 (VIR det. CCT).

CHRYSANTHEMUM L.

Chrysanthemum segetum L. Corn Marigold

A weed of cultivated ground on sandy soils. Rare in districts 2, 3, 5 and 8 but quite frequent on the Yeovil Sands of district 4. A rare casual of roadsides and waste places in the north of the county (9 & 10).

LEUCANTHEMUM Mill.

Leucanthemum vulgare Lam. Ox-eye Daisy

Chrysanthemum leucanthemum L.

In grassy places. Very common throughout the county particularly on calcareous soils.

Leucanthemum maximum (Ramond) DC. Shasta Daisy

Chrysanthemum maximum Ramond

A garden escape frequent on rubbish tips.

TANACETUM L.

Tanacetum parthenium (L.) Schultz Bip. Feverfew

Chrysanthemum parthenium (L.) Bernh.; *Matricaria parthenium* L.

A garden escape long established on walls and less permanently on roadsides and in waste places. Common near houses in all districts.

Tanacetum vulgare L. Tansy

Chrysanthemum vulgare (L.) Bernh.

On river and railway banks, by roadsides and in waste places. Common by the larger rivers and frequent elsewhere in all districts except 7 but in many places it originated as an escape from cultivation.

COMPOSITAE

ARTEMISIA L.

Artemisia vulgaris L. — Mugwort

In waste places and on roadsides. Common in all districts except in the extreme west. Not recorded in district 1.

Artemisia biennis Willd.

A rare casual first definitely identified in the county near the Avon at Bath (10) in 1925 (JWW) and still in the same area in 1965 (RGR). Also recorded at Taunton (3), 1956 (CMA), Yeovil (4), ca. 1959 (VIR), Bedminster (10), 1932 (CIS), and by the Chew Lake (10), 1961 (I. I. Jefferies).

Artemisia absinthium L. — Wormwood

In rocky grassland, on roadsides and in waste places. Long known and possibly native near the sea at Bossington (2), Kewstoke and Weston-in-Gordano (9). An uncommon but quite persistent escape from cultivation elsewhere and recorded in districts 3, 5, 7, 8 & 10.

Artemisia maritima L. — Sea Wormwood

In salt marshes on the coast and by the tidal estuaries. Locally plentiful at Porlock Weir and between Lilstock (2) and Pill (10).

CARLINA L.

Carlina vulgaris L. — Carline Thistle

Dry grassland on calcareous soils. Frequent and recorded in all districts but not recently in 1.

ARCTIUM L.

Arctium lappa L. — Greater Burdock

A. majus Bernh.

On roadsides and in waste places. Common on the Levels and near the main rivers, and frequent near the coast. Not recorded in district 1 nor recently in 7.

Arctium pubens Bab. — Burdock

A. vulgare auct.

The commonest form of Burdock on waysides and in waste places throughout the county. It is possible this species may be derived from hybridization between *A. lappa* and *A. minus*. Many intermediate forms occur and these are probably back crosses with one or other of these species.

Arctium minus Bernh. Lesser Burdock

In woodland and adjoining lanes and field borders. Common in suitable habitats in all districts.

CARDUUS L.

Carduus tenuiflorus Curt. Slender Thistle

In sandy and rocky places near the sea. Local. Near Minehead, from Stolford (2) to Highbridge (8), Brean Down and Sand Point (9). It was formerly more widespread along the coast. First reported on Steep Holm in 1952 (W. R. Price) it is now plentiful on the rocky slopes and cliffs. Very rare and only a casual inland. The only recent reports are at Bishop's Lydeard (3), 1974 (HWB), and in a disused quarry at Cross (9), 1953 (C & NS) and still there in 1960 (RB).

Carduus nutans L. Musk Thistle

In dry grassland on limestone downs and in waste places. Common on Mendip and quite frequent and widely distributed elsewhere but absent from Exmoor and the other acid moorland in the west. Not recorded in district 1.

Carduus acanthoides L. Welted Thistle

C. crispus auct.

Hedgebanks and roadsides. Common on the Lias and Oolite and frequent elsewhere. Recorded in all districts.

× **nutans** (*C.* × *orthocephalus* Wallr.) Said to be the most frequently occurring thistle hybrid and has been reported from districts 2, 4, 5 and 9 but it has not been identified in the county recently.

CIRSIUM Mill.

Cirsium eriophorum (L.) Scop. Woolly Thistle

Carduus eriophorus L.; *Cnicus eriophorus* (L.) Roth

COMPOSITAE

In dry grassland on calcareous soils. Common on the Lias and Oolite inland, less so on the coast and on the other limestones. It is particularly plentiful round Bath (10) and the first British record was from Chew Magna (10) not far away (Lobelius, 1570). Not recorded in districts 1 or 6.

× **vulgaris** (*C.* × *gerhardtii* Schultz Bip.). Although the parent species often grow together and flower at the same time hybrids are rare but have been reported from V.C.6.

Cirsium vulgare (Savi) Ten. Spear Thistle

Carduus lanceolatus L.; *Cnicus lanceolatus* (L.) Willd.

Grassland, roadsides and waste places. Very common everywhere.

Cirsium palustre (L.) Scop. Marsh Thistle

Carduus palustris L.; *Cnicus palustris* (L.) Willd.

In marshes and damp grassland. Very common in all districts particularly on acid soils, but it sometimes occurs on dry hillsides also. The white-flowered form is frequent.

Cirsium arvense (L.) Scop. Creeping Thistle

Carduus arvensis (L.) Hill; *Cnicus arvensis* (L.) Roth

Pastures, cultivated land and waste places. An abundant and troublesome weed everywhere. Forms with undulate and weakly spinous leaves are known as var. **mite** Wimm. & Grab., and are not infrequent. Plants with flat, setose-fringed leaves (var. **setosum** C. A. Mey) and those with leaves densely white-felted beneath (var. **incanum** (Fisch.) Ledeb.) are probably alien casuals. Recent records of the former were at Newbridge, Bath (10), 1974 (Dr P. Macpherson) and of the latter at Watchet Docks (2), 1971 (HWB).

× **palustris** (*C.* × *celakovskianum* Knaf) has been recorded in both V.Cs 5 & 6 but not identified recently.

(× **dissectum** was recorded by Murray from Shapwick Moor (8); this was probably an error for *C. dissectum* × *palustre)*.

Cirsium acaule Scop. Dwarf Thistle

Carduus acaulos L.; *Cnicus acaulos* (L.) Willd.

In grassland on calcareous soils. Common over much of the county but very rare in the west except on the coastal Lias. Forma **caulescens** Reichb.

Perfoliate Penny-cress, *Thlaspi perfoliatum,* Charlton Adam.
Photo: P. Wakely

Alpine Penny-cress. *Thlaspi alpestre,* Charterhouse.
Photo: P. Wakely

Wall Whitlow-grass, *Draba muralis,* Nr. Emborough.   Photo: P. WAKELY

Pale Dog-violet, *Viola lactea.*
Photo: P. Wakely

White Rock-rose, *Helianthemum apenninum,* Brean Down.
Photo: P. WAKELY

Marsh Mallow, *Althaea officinalis,* Middlezoy. Photo: P. Wakely

Rock Stonecrop, *Sedum forsteranum,* Cheddar Gorge.
Photo: P. WAKELY

Honewort, *Trinia glauca,* Uphill.
Photo: P. WAKELY

Cornish Moneywort, *Sibthorpia europaea,* Leighland Chapel. Photo: P. WAKELY

Wild Madder, *Rubia peregrina,* Thurlbear Wood.
Photo: P. WAKELY

Round-headed Club-rush, *Scirpus holoschoenus,* Berrow Dunes.
Photo: P. WAKELY

Starved Wood-sedge, *Carex depauperata*, near Cheddar.
Photo: P. Wakely

Fingered Sedge, *Carex digitata,* Avon Gorge.   Photo: P. Wakely

Somerset Hair-grass, *Koeleria vallesiana,* Uphill. Photo: P. Wakely

Bulbous Foxtail, *Alopecurus bulbosus*. Combwich
Photo: P. Wakely

Bermuda Grass, *Cynodon dactylon,* Weston-super-Mare.
Photo: P. Wakely

is quite frequent especially since the reduction in the rabbit population allowed downland grass to grow taller.

× **vulgare** (*C.* × *sabaudum* Löhr) has occurred in V.C.6 but has not been reported recently.

× **palustre** (*C* × *kirschlegeri* Schultz Bip.) Specimens collected near Wells (8) in 1851 by Babington and at Failand (10) in 1914 by Miss I. M. Roper were identified as this hybrid by Dr W. A. Sledge in 1942.

Cirsium dissectum (L.) Hill Meadow Thistle

Carduus pratensis Huds.; *Cnicus pratensis* (Huds.) Willd.

In boggy and peaty meadows and commons. Locally common near Brushford (1), on the Blackdown Hills (3, 4 & 6), on West Sedgemoor (3), the peat moors (8) and the Oxford Clay east of Bruton (7, 8 & 10). Rare elsewhere. Recorded in all districts but not recently in 2.

× **palustre** (*C.* × *forsteri* (Sm.) Loud.) is said to be frequent where the parent species grow together and has been recorded on the peat moors (8) and in other places but not recently.

GALACTITES Moench

Galactites tomentosa Moench

A casual from the Mediterranean region recorded at Dunster (2), 1978 (CAH det. BM).

SILYBUM Adans.

Silybum marianum (L.) Gaertn. Milk Thistle

Mariana lactea Hill

An introduced weed of waste places. A rare casual inland but more frequent near the sea. It is particularly persistent at Brean Down (9) where it was first recorded by St Brody (1856) and it is still abundant there in some years. Not recorded in districts 1 or 6 and not recently in 4 or 10.

ONOPORDUM L.

Onopordum acanthium L. Cotton Thistle

Waste places and on roadsides. Usually a garden escape in Somerset but possibly native at Berrow (9), where it was first collected by T. Clark in 1834, and at South Cadbury (5). Recorded in all districts except 1, 6 and 7.

COMPOSITAE

CENTAUREA L.

Centaurea scabiosa L. Greater Knapweed

Dry grassland and hedgebanks, mainly on calcareous soils. Common on the limestones but rare elsewhere. Recorded in all districts but very scarce in district 1.

Centaurea montana L.

Garden escape.

Centaurea cyanus L. Cornflower

Formerly a cornfield weed but apparently never as common in Somerset as in some other parts of the country. It was already rare by the end of the nineteenth century and has only been recorded a few times since. The only recent record in a cornfield was at Buckland St Mary (6), 1968 (NGC), but it is occasionally seen as a garden escape or as a casual on rubbish tips.

Centaurea nigra L. sensu lato Common Knapweed

Grassland, roadsides and waste places. Very common throughout the county. The aggregate is usually divided in Britain into two species or subspecies, **C. debeauxii** Gren. & Godr. ssp. **nemoralis** (Jordan) Dostál (*C. nemoralis* Jord.) and **C. nigra** sens. strict., but the two taxa apparently hybridize freely and so many intermediates occur it is impracticable to separate them satisfactorily or to work out their respective distributions.

†**Centaurea calcitrapa** L. Red Star Thistle

A rare casual, at one time persistent near Ham Hill (4) and thought to be possibly native there but not recorded since 1883 nor anywhere else in the county since 1926.

Centaurea diluta Aiton

A bird-seed alien increasingly found on rubbish tips, first recorded in the county in 1966 at Henstridge (7) (CAH).

SERRATULA L.

Serratula tinctoria L. Saw-wort

In damp meadows and on commons and hedge and railway banks. Frequent on the Blackdown Hills (3, 4 & 6) and on the Oxford Clay near the eastern borders (7, 8 & 10). Rather rare in the west and on more calcareous soils. Not recorded in district 2.

CARTHAMUS L.

Carthamus tinctorius L.

A casual of rubbish tips first recorded in the county at St Anne's, Brislington (10), 1917 (CIS).

CICHORIUM L.

Cichorium intybus L. Chicory

Roadsides, farmyards and waste places. Sometimes an escape from cultivation. A colonizer of new roadworks. Frequent and widespread. Recorded in all districts except 7 and not recently in 1.

LAPSANA L.

Lapsana communis L. Nipplewort

A weed of cultivated ground, roadsides and waste places. Very common throughout the county.

ARNOSERIS Gaertn.

†**Arnoseris minima** (L.) Schweigg. & Koerte Lamb's Succory

A. pusilla Gaertn.

Recorded for Somerset in the *New Botanist's Guide* (1835) on the authority of Gapper but not seen since.

HYPOCHOERIS L.

Hypochoeris radicata L. Cat's-ear

A weed of grassland, very common in all districts.

†**Hypochoeris glabra** L. Smooth Cat's-ear

A plant of grassy places on sandy soils, formerly recorded at Bossington and Minehead Warren (2), Berrow, Brean Down and Kewstoke (9). Reported in 1932 by W. D. Miller to be still frequent on Minehead Warren but its presence there has not been confirmed since 1956. In 1946 Mrs. C. I. and N. Y. Sandwith, having failed to find the plant in any of the North Somerset localities in spite of repeated searches, began to wonder if it really occurred in the area. They were unable to trace any North Somerset specimens in any of the main herbaria, including those of Druce, White, Marshall and Miss Roper, supporting the records and they concluded they were probably based on misidentifications. It seems more likely however to be a case of the final extinction of a rare and declining species.

LEONTODON L.

Leontodon autumnalis L. Autumn Hawkbit

Meadows, pastures and roadsides. Very common in all districts.

Leontodon hispidus L. Rough Hawkbit

In dry grassland. Very common everywhere, particularly on calcareous soils.

Leontodon taraxacoides (Vill.) Mérat Lesser Hawkbit

L. leysseri G. Beck; *L. hirtus* auct.; *Thrincia hirta* Roth

Dry grassland. Common and widespread.

PICRIS L.

Picris echioides L. Bristly Oxtongue

Helmintia echioides (L.) Gaertn.

Roadsides and waste places especially on clay soils. Common throughout the lowlands and recorded in all districts except 1.

Picris hieracioides L. Hawkweed Oxtongue

On dry banks on calcareous soils. Common on the Lias and Oolite, less so elsewhere but recorded in all districts except 1.

TRAGOPOGON L.

Tragopogon pratensis L. Goat's-beard

Grassy and waste places. Very common throughout the county except on Exmoor and other high ground. Most plants found are ssp. **minor** (Mill.) Wahlenb. (*T. minor* Mill.); ssp. **pratensis** is decidedly rare.

Tragopogon porrifolius L. Salsify

An escape from cultivation which has been long established on the Carboniferous Limestone at Brean Down and Weston-super-Mare (9). Recorded as a casual in a few other places where it has not persisted.

LACTUCA L.

Lactuca serriola L. Prickly Lettuce

Roadsides, waste places and rubbish tips. First definitely recorded in the county in 1934 at Taunton (3) (WW), it remained a rare casual until about 1944 when it became quite frequent on bombed sites about Bristol and Bath (10). It was reported at Bridgwater (3 & 5) in 1955 but it was another fifteen years before its spread became more rapid. It is now plentiful where large amounts of earth have been moved, as on new housing estates and roadworks and on rubbish tips particularly near the larger towns, e.g., Taunton (3), Bridgwater (5), Highbridge (8), Clevedon and Portishead (9), Bristol and Bath (10). It also occurs as a roadside weed and in waste places on the Lias in districts 2 and 5, and is now recorded in all districts except 1 and 6. Most plants in Britain have unlobed leaves (forma **integrifolia** (S. F. Gray) S. D. Prince & R. N. Carter or var. **dubia** (Jord.) Rouy) a feature which is not mentioned in most Floras and this has caused much confusion with *L. virosa,* so the history of this species in Somerset may go back to 1907 or even earlier.

†**Lactuca virosa** L. Great Lettuce

Never more than a rare casual in Somerset there is some doubt whether this species has ever occurred in the county; there are certainly no reliable recent records. Credited to Somerset in the *New Botanist's Guide* (1835) and in *Topographical Botany* (1873-74), there were no localized records until 1900 [Pill (10) (JWW)] and 1907 [railway near Langport (3)(ESM)]. A few later records were almost certainly errors for the last species, as these earlier ones may well have been, as many botanists have recorded the form of *L. serriola* with unlobed leaves as *L. virosa.*

MYCELIS Cass.

Mycelis muralis (L.) Dumort. Wall Lettuce

On old walls, in woodland and on rocky banks in hilly districts. Frequent on Exmoor, the Quantocks, Mendip and other Carboniferous Limestone in North Somerset, and on the Oolite from Bruton (8) to the Bath (10) area. Absent from much of the lowlands. Not recorded recently in district 4 and very rare in districts 5 and 7.

SONCHUS L.

Sonchus arvensis L. Corn Sow-thistle

A weed of cornfields and other arable land and waste places. Very common in all districts.

Sonchus oleraceus L. Smooth Sow-thistle

Arable land, gardens and waste places. Very common throughout the county.

Sonchus asper (L.) Hill Prickly Sow-thistle

Cultivated land, waste places and roadsides. Abundant everywhere and the commonest species of the genus in Somerset.

CICERBITA Wallr.

Cicerbita macrophylla (Willd.) Wallr. Blue Sow-thistle

A garden escape increasingly established on roadside verges. First noticed in the wild in Somerset at Ston Easton (10), 1956 (R & IR), it has now been recorded at Trent (5) (Dorset), 1969 (CAH), and in six more localities in district 10.

HIERACIUM L. (Hawkweed)

Sect. **Amplexicaulia** Zahn

Hieracium speluncarum Arv.-Touv.

Misidentified at different times as *H. amplexicaule* L. and *H. pulmonarioides* Vill., this introduced species is established on old walls at Mells (10) where it was first reported in 1913.

Sect. **Oreadea** Zahn

Hieracium schmidtii Tausch

H. lima F. J. Hanb.

Only in the Cheddar Gorge (9).

Hieracium cyathis (A. Ley) W. R. Linton

Limestone cliffs in the Cheddar Gorge (9).

Hieracium eustomon (E. F. Linton) Roffey

A rare and local species of the rocky coast between the Devon border and Minehead (2) which has not been positively identified recently.

Hieracium stenolepiforme (Pugsl.) Sell & West

Endemic to the Cheddar Gorge (9).

Hieracium angustisquamum (Pugsl.) Pugsl.

Another of the rare Cheddar Gorge (9) hawkweeds.

Sect. **Vulgata** F. N. Williams

Hieracium grandidens Dahlst.

An introduced species first recorded on railway banks near the Devon border east of East Anstey (1) in 1916 (ESM) and now quite frequent in that area.

Hieracium sublepistoides (Zahn) Druce

Another introduced species recorded in an old quarry in Murder Combe. near Mells (10), in 1975 (Dr B. N. K. Davis). A plant from a field bank nearer Mells found by C. A. Howe in 1968 was determined by P. D. Sell as *H. exotericum* agg. and may well have been the same species.

Hieracium stenstroemii (Dahlst.) K. Joh.

H. glevense (Pugsl.) Sell & West

A species native in rocky areas of south-east Wales and adjacent parts of England, once recorded on the towpath under Leigh Woods (10), 1954 (B. Miles det. Sell & West).

Hieracium vulgatum Fr.

A northern species often erroneously recorded in the county but the only confirmed occurrence, presumably a casual one, was by the railway track near Bruton (8), 1966 (CAH det. Sell).

Hieracium maculatum Sm.

Quarries, old walls, colliery tips, railway ground and very occasionally in plantations. Believed to be originally an introduction because of its predilection for man-made habitats, it was well established about Bath (10) by 1866 according to Jenyns and is now quite common in the area between Wells (8), Frome and Bath (10). It also occurs in some of the limestone quarries of district 9. Outside these areas it is rare and almost confined to railway ground in a few spots in districts 3, 5 and 7.

Hieracium subamplifolium (Zahn) Roffey

An endemic species of South Wales probably introduced in Somerset. It was first detected in the county in 1965 at Charlton Mackrell (5) (CAH det. Sell) and has since been recorded at a number of places near the railway between Somerton (5) and Bruton (8), and at Shipham Hill quarry (9).

Hieracium acuminatum Jord.

H. strumosum (W. R. Linton) A. Ley; *H. lachenalii* auct.

Rocks, banks and walls. The most widespread Hawkweed in the county, quite frequent in the north and west and recorded in all districts except 4 and 6. The closely allied **H. cheriense** Jord. ex Bor. also occurs in the extreme west (1 & 2) but has not been identified recently by an expert.

Hieracium diaphanum Fr.

inc. *H. anglorum* (A. Ley) Pugsl.

Common further north but rare in southern England. There have been a few reports from Somerset but the only reliable recent one was from an old quarry between Shipham and Cheddar (9), 1975 (B. N. K. Davis det. Sell).

Sect. **Tridentata** F. N. Williams

Hieracium trichocaulon (Dahlst.) Johans.

Grassy roadsides and railway banks. Rather rare and recorded recently in only a few scattered localities in districts 1, 2, 3, 8 and 10. Some of these records may be of the closely allied **H. calcaricola** (F. J. Hanb.) Roffey. Under the aggregate name *H. tridentatum* Hawkweeds of this group were also recorded formerly in districts 4 and 7.

Sect. **Umbellata** F. N. Williams

Hieracium umbellatum L.
ssp. **umbellatum**

Heaths and shady banks, mainly on acid soils. Frequent on the Blackdown Hills (3, 4 & 6) and the peat moors (8) but rather thinly scattered elsewhere in the west (2 & 3) and towards the eastern border (7, 8 & 10). Not recorded in district 5 nor recently in 1 or 9.

ssp. **bichlorophyllum** (Druce & Zahn) Sell & West

H. bichlorophyllum (Druce & Zahn) Pugsl.

In woods and on shady banks. Frequent in the west in districts 1 and 2. Only once recorded in district 3; a specimen collected in 1825 near Taunton by A. & J. S. Henslow is in the Cambridge University herbarium. Where the distributions of this and ssp. *umbellatum* overlap in district 2 intermediates occur.

Sect. **Sabauda** F. N. Williams

Hieracium sabaudum L.

H. perpropinquum (Zahn) Duce

Open woodlands, heaths, roadsides and railway banks. Frequent in the west (1, 2 & 3) and in districts 7 and 10. Elsewhere the only recent records were at Trent (5) (Dorset), 1968 (CAH), East Lydford (5), 1969 (CAH) and Wambrook (6) (V.C.9), 1967 (NGC).

Hieracium rigens Jord.

A species of south-eastern England found in 1980 by D. E. Green on the track of a long-disused railway near Paulton (10) and later the same year by the railway at Newton St Loe (10) by R. D. Randall. Specimens from both localities were determined by C. E. A. Andrews.

Hieracium salticola (Sudre) Sell & West

First recorded in Somerset in Leigh Woods (10), 1955 (A. H. G. Alston det. Sell & West) and later by a disused railway at Bath (10), 1980 (DG det. Andrews).

Hieracium vagum Jord.

Native further north but an introduced species in Somerset where it has been recorded on railway banks at Marston Magna (5), 1966 (CAH det.

Sell), Saltford (10), 1977 (ALG det. Sell), Radstock (10) and Bath (10), 1980 (DG det. Andrews).

Subgen. **Pilosella** (Hill) S. F. Gray

Hieracium pilosella L. Mouse-ear Hawkweed

Dry grassy banks, downland, rocks and wall tops. Very common throughout the county.

Hieracium aurantiacum L. ssp. **carpathicola** Naeg. & Peter

H. brunneocroceum Pugsl.

A garden escape widely naturalized on walls and rocky banks near houses and recorded in all districts.

× **pilosella** (*H.* × *stoloniflorum* Waldst. & Kit.) Garden escape recorded on a wall at Maperton (5), 1965 (CAH).

Crepis L.

Crepis vesicaria L. ssp. **haenseleri** (Boiss. ex DC.) P. D. Sell Beaked Hawk's-beard

C. vesicaria ssp. *taraxacifolia* (Thuill.) Thell.; *C. taraxacifolia* Thuill.

Roadsides, railway banks, cultivated and waste land. The earliest records in Somerset were dated 1883 and it was still rare at the end of the century but began to spread soon afterwards. It is now very common over most of the county except on acid soils. Recorded in all districts but still rare in district 1.

Crepis setosa Haller f. Bristly Hawk's-beard

A rare casual which appeared in a newly-laid garden in Bath (10), 1979 (HGW).

Crepis biennis L. Rough Hawk's-beard

In pastures and on roadsides, probably introduced with grass seed. First recorded in the county in 1883 in a field near Cheddar (9) at the same time as *C. vesicaria* and both Murray and White expected it to spread like that species but it did not do so except locally. It persisted there until the fields were ploughed up in 1908 but it has not reappeared. In 1962 it was reported at Swainswick and Batheaston (10) and it is still to be found in the area. In several other localities in districts 9 and 10 it soon died out.

Crepis capillaris (L.) Wallr. Smooth Hawk's-beard

C. virens L.

Grassland, cultivated and waste land. A very common and very variable weed recorded in all districts.

TARAXACUM Weber (Dandelion)

Continental botanists have long divided this genus into a large number of microspecies while in Britain it has usually been the practice to treat it as a few aggregate species. Work by A. J. Richards and the publication of his *Taraxacum Flora of the British Isles* in 1972 has generated a new interest in the genus. The following microspecies are credited to Somerset by Richards in his book or have been identified by him since.

Sect. **Erythrosperma** (H. Lindb. f.) Dahlst. (*T. laevigatum* agg.)

Dry rocky and sandy places. Common near the sea, on Mendip and on other calcareous downs.

T. brachyglossum (Dahlst.) Dahlst. V.Cs.5 & 6
T. lacistophyllum (Dahlst.) Raunk. V.C.6
T. rubicundum (Dahlst.) Dahlst. V.C.6.
T. laetum (Dahlst.) Dahlst. V.C.6. Berrow (9), 1975 (JGK).
T. fulvum Raunk. V.C.6.
T. fulviforme Dahlst. V.C.6.
T. oxoniense Dahlst. V.Cs. 5 & 6.
T. glauciniforme Dahlst. V.C.6.
T. proximum (Dahlst.) Dahlst. V.C.6.
T. simile Raunk. V.C.6. Steep Holm, 1977 (JGK).

Sect. **Obliqua** Dahlst. (inc. in *T. laevigatum* agg.)

T. platyglossum Raunk. V.C.6.

Sect. **Spectabilia** Dahlst. (*T. spectabile* agg.)

In wet and marshy places. Rather rare and apparently less frequent than formerly. Old records for *T. paludosum* and *T. palustre* probably belong here rather than to Sect. **Palustria** which does not apparently occur in Somerset.

T. unguilobum Dahlst. V.C.6.
T. spectabile Dahlst. V.C.6. Witham (10), 1972 (CAH).
T. praestans H. Lindb. f. V.C.5. V.C.6. Ashcott (8), 1975 (JGK).
T. nordstedtii Dahlst. V.C.5. V.C.6. Penselwood (7), 1968 (CAH).

T. adamii Claire V.C.5 Britty Common (4), 1974 (JGK). V.C.6.

Sect. **Vulgaria** Dahlst. (*T. officinale* agg.)

Roadsides, grassland, cultivated and waste ground. Very common everywhere.

T. subcyanolepis M. P. Chr. V.C.5. Thornfalcon (3), 1978 (MMcCW).
T. linguatum Dahlst. ex M. P. Chr. & Wünst. V.C.5. Thurlbear (3), 1978 (MMcCW).
T. alatum H. Lindb. f. V.C.6.
T. croceiflorum Dahlst. V.C.5. Buckland St Mary (6), 1978 (MMcCW).
T. expallidiforme Dahlst. V.C.5. Crewkerne (4), 1975 (JGK). V.C.6.
T. ekmanii Dahlst. V.C.5. Thornfalcon (3), 1978 (MMcCW).
T. porrectidens Dahlst. V.C.5. Britty Common (4), 1974 (JGK).
T. pectinatiforme H. Lindb. f. V.C.5. V.C.6. Shapwick (8), 1976 (EC).
T. cordatum Palmgr. V.C.5. Thornfalcon (3), 1978 (MMcCW). V.C.6. Berrow (9), 1975 (JGK).
T. longisquameum H. Lindb.f. V.C.5. Thurlbear (3), 1978 (MMcCW).
T. dahlstedtii H. Lindb. f. V.C.5. Dunster (2), 1978 (MMcCW).
T. christiansenii Hagl. V.C.5. Chillington (4), 1974 (JGK).
T. hamatum Raunk. V.C.5. Crewkerne (4), 1975 (JGK). V.C.6. Downhead (10) 1972 (CAH).
T. hamatiforme Dahlst. V.C.5. North Curry (3), 1975 (JGK).
T. marklundii Palmgr. V.C.5. Britty Common (4), 1974 (JGK).
T. oblongatum Dahlst. V.C.6. Shapwick (8), 1976 (EC).
T. reflexilobum H. Lindb. f. V.C.6 Steep Holm, 1977 (JGK).
T. crispifolium H. Lindb. f. V.C.5. Purtington (4), 1974 (JGK).
T. tarachodum Haglund, van Soest. & Zahn V.C.6 Dunster (2), 1978 (MMcCW).

MONOCOTYLEDONES

ALISMATACEAE

Baldellia Parl.

Baldellia ranunculoides (L.) Parl. Lesser Water-plantain

Alisma ranunculoides L.

In peaty ditches. Always rare and local and now only to be found on Northmoor (3), King's Sedgemoor (5), the peat moors (8) and in the Gordano valley (9), although formerly more widespread in the lowlands in district 9 with some outlying localities in districts 4 and 10.

ALISMA L.

Alisma plantago-aquatica L. Common Water-plantain

In moor ditches, by ponds, streams and rivers. Very common throughout the lowlands and frequent elsewhere except in the west. Only once recorded in district 1 (Upton, 1968 (CAH)) and local in district 2.

Alisma lanceolatum With. Narrow-leaved Water-plantain

In ditches, ponds and canals. Distribution rather uncertain because of confusion with *A. plantago-aquatica* from which it was not distinguished by many botanists until fairly recently. Frequent in the central lowlands (3, 4, 5 & 8). Rare outside this area but recorded in a few places in districts 9 and 10 with isolated former records in 2 and 6.

SAGITTARIA L.

Sagittaria sagittifolia L. Arrow-head

In ditches, ponds, canals, streams and rivers. Common in the central and northern lowlands (3, 4, 5, 8 & 9) but not seen recently in the Gordano valley (9). Outside this area it occurs only in ditches near Stolford (2), in the Rivers Cale (7), Frome and Avon (10) and in the Kennet & Avon Canal (10).

BUTOMACEAE

BUTOMUS L.

Butomus umbellatus L. Flowering Rush

In ditches, rivers, canals, ponds and reservoirs. Formerly common throughout the lowlands and still quite frequent there (3, 4, 5, 8 & 9). Also recorded recently near Minehead (2), in the Rivers Cale and Stour (7), in ponds at Redlynch (8) and Ham Green (10) and in the Blagdon reservoir (9). Formerly frequent in the River Avon it has not been seen there for many years but it still occurs in the Kennet & Avon Canal (10).

HYDROCHARITACEAE

HYDROCHARIS L.

Hydrocharis morsus-ranae L. Frog-bit

In ditches and rhines. Wall Common (2), 1968 (HWB), and abundant throughout the Levels (3, 4, 5, 8 & 9). The only record outside this area and above the 50-foot contour was in a pool on Mendip at about 800 feet above Westbury (9), 1970 (JA).

STRATIOTES L.

†**Stratiotes aloides** L. Water Soldier

Only a casual introduction in Somerset. It appeared in abundance in two ditches on South Moor, Glastonbury (8) in 1963 (AFD) and persisted there until the drought of 1976 when these ditches completely dried out. It was also known for a time near Yatton (9) but had disappeared by 1950.

ELODEA Michx

Elodea canadensis Michx Canadian Waterweed

Anacharis alsinastrum Bab.

Ditches, canals, streams and ponds. An American plant, introduced to Britain in 1842 and first noticed in Somerset about 1856, which spread rapidly but began to decrease again by the end of the century. It is still common throughout the Levels and quite frequent elsewhere in the county except in the west (1 & 2) where it never became as common as in other parts and is now rather rare.

Elodea ernstae St John

E. callitrichoides auct.

An aquarist's throw-out recorded in the Kennet & Avon Canal (10) in 1956 (JPMB) and again in 1961 (PJMN), and in the King's Sedgemoor Drain at Bawdrip (5) in 1975 (JA).

A third species of the genus, **E. nuttallii** (Planch.) St John, was first identified in Britain in 1974 and should be looked for as it is spreading rapidly in other parts of the country.

JUNCAGINACEAE

TRIGLOCHIN L.

Triglochin palustris L. Marsh Arrowgrass

Damp pastures, marshes and ditchbanks. Not common but widespread and recorded in all districts. Not seen recently in district 7.

Triglochin maritima L. Sea Arrowgrass

Salt marshes on the coast and up the tidal river estuaries. Rather scarce in the west from Porlock to Hinkley Point (2) but common from Stolford (2) to Uphill (9) and from Kewstoke (9) to the Avon banks below Leigh Woods (10).

APONOGETONACEAE

APONOGETON L. f.

Aponogeton distachyos L. f.

An aquarist's throw-out.

ZOSTERACEAE

ZOSTERA L.

Zostera marina L. Eel-grass

In the sea near the low-water mark. Very rare and decreasing. It has only been reported in recent years from near Stolford (2) (1957 and 1969). It was formerly recorded at intervals between the mouth of the Brue (8) and Brean Down (9) but it has not been reported from this stretch of coast since 1929.

†**Zostera angustifolia** (Hornem.) Reichb. Narrow-leaved Eel-grass

In the sea near low-water mark. Only once recorded in the county, in Kewstoke Bay (9), 1916 (CIS & IMR).

†**Zostera noltii** Hornem. Dwarf Eel-grass

Z. nana auct.

On muddy seashores. T. B. Flower told White in 1884 that he had gathered this in Somerset estuaries but the record has been doubted as White could find no specimen in Flower's herbarium. An earlier reference to North Somerset in *Topographical Botany* (1873-74) was an error due to confusion of the Devonshire Axe with the river of the same name in Somerset.

POTAMOGETONACEAE

Potamogeton L.

Potamogeton natans L. Broad-leaved Pondweed

In ponds, canals and slow streams. Rather rare but widespread and recorded in all districts.

Potamogeton polygonifolius Pourr. Bog Pondweed

In bogs, peaty ditches and streams and pools of acid water. Common on the western moors (1 & 2), Blackdown Hills (4 & 6), peat moors (8) and the Mendip sandstone (9 & 10). Very rare elsewhere.

Potamogeton coloratus Hornem. Fen Pondweed

In fen ditches. Rare and local. More plentiful at the eastern end of King's Sedgemoor (5) than anywhere else, it also occurs on the peat moors (8), Kenn Moor and in the Gordano valley (9). Formerly recorded near Axbridge (9) but not seen there for many years.

Potamogeton nodosus Poir. Loddon Pondweed

P. drucei Fryer

In the River Avon. Although a specimen was collected on the Gloucestershire side of the river opposite Brislington by the Rev W. H. Painter in 1884 this species was not recognized until 1916 when it was found at Saltford by C. Bucknall and Mrs Wedgwood. At that time it was believed to be a British endemic and was named *P. drucei* Fryer but in 1939 Dandy and Taylor pointed out in a paper in the *Journal of Botany* that it is conspecific with the European species *P. nodosus* Poir. It apparently increased considerably about 1920 and by 1940 it was abundant over the whole course of the river from Brislington to the Wiltshire border and beyond. It is still quite plentiful, particularly near Claverton (10).

Potamogeton lucens L. Shining Pondweed

Rivers and streams, rhines and ditches. Frequent in central Somerset in districts 3, 4, 5 & 8. Elsewhere the only recent records are from the Cheddar reservoir (9), 1979 (RSC), Nailsea Moor (9), 1978 (PJMN) and in the Avon below Bath (10). Formerly it was also known on Kenn Moor (9) and in the River Frome (10).

× **perfoliatus** (*P.* × *salicifolius* Wolfg.; *P. decipiens* Nolte). In rivers and canals. Recorded formerly in the Brue (8) and in the Avon and the canals near Bath (10) but not seen in the county since 1940.

†**Potamogeton gramineus** L. Various-leaved Pondweed

P. heterophyllus Schreb.

In pools and ditches. Old records for Weston-super-Mare (9) and near Bristol and Bath (10) were considered misnomers by Murray but there are confirmed records by White from the Gordano valley (9) and disused canal at Paulton (10), none later than 1904.

†**Potamogeton alpinus** Balb. Red Pondweed

Streams and ditches. Recorded at Wembdon (3) by J. C. Collins, in the Cary near Somerton (5) by J. G. Baker and near Axbridge (9) by White but not seen anywhere in the county since 1896.

Potamogeton perfoliatus L. Perfoliate Pondweed

Rivers, streams, canals and reservoirs, in deepish water. Rare and decreasing. Recorded recently only in the Bridgwater & Taunton canal at North Newton (3), 1972 (JGK), the upper reaches of the Parrett (4), 1969 (JGK), the Axe near Chard (6), 1976 (JGK), the Blagdon reservoir (9), 1975 (JA) and the By Brook near Bathford (10), 1962 (R & IR). Formerly more widespread and recorded in districts 5 and 8 also.

Potamogeton friesii Rupr. Flat-stalked Pondweed

In canals and rivers. Very rare and now only in the Kennet & Avon canal near Bath (10). Formerly also in the old Coal Canal, which is now largely filled in, and recorded in the Avon in 1917 (IMR). Only once recorded in South Somerset, in the Bridgwater & Taunton canal (3), before 1914 (HST).

Potamogeton pusillus L. Lesser Pondweed

P. panormitanus Biv.

In ponds, canals and moor ditches. Frequent in the lowlands of districts 3, 4, 5, 8 and 9. Its former distribution is difficult to determine because of confusion until recently with *P. berchtoldii* to which many of the older records belong.

POTAMOGETONACEAE

Potamogeton obtusifolius Mert. & Koch — Blunt-leaved Pondweed

Ponds and lakes. Very rare. First recorded in the county in a pond at Norton Fitzwarren (3), 1973 (HWB conf. J. E. Dandy), and since found in a lake at Otterford (6), 1978 (JGK conf. A. C. Jermy).

Potamogeton berchtoldii Fieb. — Small Pondweed

P. pusillus auct.

In ponds and ditches. Frequent and widespread. More plentiful in the lowlands than elsewhere but recorded in all districts except 1.

Potamogeton trichoides Cham. & Schlecht. — Hair-like Pondweed

Canals, ditches and ponds. Rare. First detected in the county in Blagdon Lake (9), 1934 (Miss E. Claydon conf. J. E. Dandy), but not seen again until 1972 when it was found in several places in the lowlands of districts 3, 4 and 5 (JGK all conf. J. E. Dandy).

Potamogeton crispus L. — Curled Pondweed

Ditches, rivers and ponds. Common throughout the lowlands and frequent elsewhere except in the west where it is rare. Recorded in all districts.

Potamogeton pectinatus L. — Fennel Pondweed

inc. *P. flabellatus* Bab.

Rivers, ditches, canals and ponds. Common in the lowlands and in some rivers elsewhere but rare in the west. Recorded in all districts.

GROENLANDIA Gay

Groenlandia densa (L.) Fourr. — Opposite-leaved Pondweed

Potamogeton densus L.

Rhines, ditches and ponds. Local and decreasing. Formerly rather common and widespread in the lowlands but not seen recently in districts 7 and 10. Absent from districts 1 and 2 and never recorded in the county west of Taunton (3).

RUPPIACEAE

RUPPIA L.

Ruppia cirrhosa (Petagna) Grande — Spiral Tasselweed

R. spiralis L. ex Dumort.; *R.maritima* auct.

In brackish pools. Very rare and only once recorded recently, in the marshland between Brean Down and Uphill (9), 1967 (PJMN), where it was first reported by Banks and Lightfoot in 1773. Specimens collected near Kingston Seymour (9) in 1920 by Miss I. M. Roper are in the herbaria at Taunton and at Leicester University.

Ruppia maritima L. — Beaked Tasselweed

R. rostellata Koch

Brackish pools. Rare. Only seen recently near Stolford (2) and Portbury (10) but formerly recorded near Porlock and Minehead (2) also.

ZANNICHELLIACEAE

ZANNICHELLIA L.

Zannichellia palustris L. — Horned Pondweed

In ditches, ponds and streams. Formerly common and widespread but now decreasing. Still quite frequent throughout the Levels though rarer elsewhere. Not recorded in district 1 nor recently in 4.

LILIACEAE

NARTHECIUM Huds.

Narthecium ossifragum (L.) Huds. — Bog Asphodel

In wet moorland bogs. Common on Exmoor (1 & 2), at the north end of the Quantocks (2 & 3) and on the Blackdown Hills. Rare on the peat moors (8) and in a few spots on the Mendip sandstone (9).

CONVALLARIA L.

Convallaria majalis L. — Lily of the Valley

In rocky woods. Rare. Cheddar Wood and near Churchill (9), Leigh Woods and Asham Wood (10). Probably an introduction or a garden escape in a few other localities. In Tetton Woods (3) a pink-flowered form was reported by F. J. Hanbury in 1872 and it persisted for many years but has not been seen since 1924.

LILIACEAE

POLYGONATUM Mill.

Polygonatum odoratum (Mill.) Druce — Angular Solomon's Seal

P. officinale All.

Rocky woods on limestone. Very rare. Cheddar Gorge (9) and Leigh Woods (10). Formerly in Asham Wood (10) but not seen there since 1918. Recorded by Sole (1791) in East Harptree Wood (10) but long since gone. Other records near Oare (2), Wells (8), Congresbury (9), Mells and Babington (10) were probably introductions which have not persisted.

Polygonatum multiflorum (L.) All. — Solomon's Seal

In woods on limestone. Common on Mendip east of Wells and East Harptree and on the Oolite between Batcombe and Bath, in districts 8 and 10. Rare elsewhere as a native but widespread and recorded in all districts except 6 as an introduction or garden escape.

× **odoratum** (*P.* × *hybridum* Brügger). Widely cultivated and many of the escapes referred to above probably belong here.

ASPARAGUS L.

Asparagus officinalis L. — Wild Asparagus

A bird-sown garden escape thoroughly naturalized on the sandy coast between Burnham (8) and Brean Down (9), where it has been known since 1836 (T. Clark). It was recorded by Banks by the river under Leigh Woods (10) in 1767 and was seen there as recently as 1914 (CIS). Only a casual and impermanent elsewhere.

RUSCUS L.

Ruscus aculeatus L. — Butcher's Broom

Hedges and wood borders. Only a planted introduction in Somerset.

LILIUM L.

Lilium martagon L. — Martagon Lily

A garden escape rarely naturalized in woods.

Lilium pyrenaicum Gouan

Another rare garden escape.

FRITILLARIA L.

†**Fritillaria meleagris** L. Fritillary

Formerly in meadows near Compton Martin, Barrow Gurney (9) and Norton St Philip (10). It is not known when it finally disappeared from Somerset but there are no records of it since the publication of White's *Bristol Flora* (1912).

TULIPA L.

Tulipa sylvestris L. Wild Tulip

An introduced plant sometimes naturalized in old orchards and meadows. The only recent record is near Bruton (8) (CAH) where it has been known for over fifty years. It was long established in fields at Combe Hay (10) but has not been seen there since 1933, and a colony in an orchard at Wheathill (5), known since 1912, has also gone. A few other records have proved less long-lasting.

GAGEA Salisb.

Gagea lutea (L.) Ker-Gawl. Yellow Star of Bethlehem

G. fascicularis Salisb.

In woods. Very rare. Collected by Lobelius in Somerset in 1570 it now survives in only one locality, near Frome (10). Formerly in several woods near Bath and in some near Bristol (10) where it was last recorded in 1925. It is still to be found near Bath just over the county borders in Wiltshire and Gloucestershire.

ORNITHOGALUM L.

Ornithogalum umbellatum L. Star of Bethlehem

Originally introduced but now thoroughly naturalized on grassy roadsides and in fields on calcareous soils. Uncommon but widespread and recorded in all districts except 1, 6 and 7 and not recently in 4.

Ornithogalum pyrenaicum L. Bath Asparagus

In woods and on roadside banks on calcareous soils. Locally abundant in the north-east of the county in district 10 from Stockwood to the Wiltshire border east of Bath and south to Beckington. Outside this area it is extremely rare though it has been reported in the past as far south as Nunney. The young spikes are no longer sold in Bath as a substitute for

Asparagus although they still were in 1912 according to White; Sole (1791) warns that "it is not very wholesome; for if eaten plentifully it occasions nausea and oppression of the breath."!

SCILLA L.

†**Scilla autumnalis** L. Autumn Squill

Formerly at Burwalls on the Somerset side of the Avon Gorge (10) but apparently eradicated there by the construction of the Suspension Bridge.

HYACINTHOIDES Medic.

Hyacinthoides non-scripta (L.) Chouard ex Rothm. Bluebell

Endymion non-scriptus (L.) Garcke; *E. nutans* Dum.; *Scilla nutans* Sm.

In woods and hedgerows. Very common throughout the county.

Hyacinthoides hispanica (Mill.) Rothm. Spanish Bluebell

Endymion hispanicus (Mill.) Chouard

Garden escape.

MUSCARI Mill.

Muscari neglectum Guss. ex Ten. Grape Hyacinth

M. atlanticum Boiss. & Reut.; *M. racemosum* auct.

Only a garden escape in Somerset.

COLCHICUM L.

Colchicum autumnale L. Meadow Saffron

Rides and wood borders and occasionally in meadows. Much less frequent in the latter than formerly due to ploughing of grassland and deliberate eradication as poisonous to livestock. Rare in South Somerset and now confined to a few spots on Quantock (3) and near Yeovil (4) and Templecombe (7). Still quite common in woods in most of North Somerset but rare near Bristol (9 & 10).

PARIS L.

Paris quadrifolia L. Herb Paris

In woods on calcareous soils. Frequent but seldom in large quantity and easily overlooked. Absent from districts 1 and 2 and only once recorded west of Wellington (3), near Milverton (3), 1958 (AD & OMH). Recorded in all other districts but not in the lowland areas nor near the coast.

JUNCACEAE

JUNCUS L.

Juncus squarrosus L. Heath Rush

Acid heaths and moors. Common on Exmoor and frequent on the Brendon (1 & 2) and Blackdown Hills (3, 4 & 6). It also occurs in a few spots on the Mendip sandstone (9 & 10) and on the Greensand at Witham Park (10), 1969 (CAH). There are old records for the Quantocks (3), peat moors (8) and Gordano valley (9) but it has not been seen recently in any of these places.

Juncus tenuis Willd. Slender Rush

J. macer Gray

Muddy roadsides and tracks. An alien first recorded in the county in 1914 in Leigh Woods (10) by Mrs. C. I. Sandwith. Still quite plentiful there but rare elsewhere and to date it has only been recorded in a few isolated places in districts 3, 6, 8 and 9.

Juncus compressus Jacq. Round-fruited Rush

Damp pastures and roadsides. Very rare and only recorded recently at East Chinnock (4), 1969 (CJC & JGK), Ashcott Heath (8), 1972 (JTK), Bruton (8), 1963 (CAH), Blagdon (9), 1975 (JA), and North Stoke (10), 1965 (RGR). Formerly rather more widespread in the north of the county and recorded in districts 5 and 6 also.

× **gerardi** (*J.* × *royeri* P. Fourn.). Plants from the salt marshes near the coast at Berrow (9) intermediate in some characters between these two species have been reported as this hybrid but they are probably abnormal forms of *J. gerardi*.

JUNCACEAE

Juncus gerardi Lois — Salt-marsh Rush

In salt marshes on the coast and up the tidal estuaries. Common from Minehead (2) to Pill (10). Formerly recorded as far west as Porlock Weir.

Juncus bufonius L. — Toad Rush

On damp open ground, on tracks and by ponds. Very common throughout the county.

Juncus foliosus Desf.

Only recently recognized in Britain as distinct from *J. bufonius,* from which it differs by its broader bright-green leaves, striped tepals and conspicuously-ribbed seeds. As yet only identified in district 9, by a rill on Blackdown, 1916 (H. F. Devis in **K**) and in a cornfield at Weston-in-Gordano, 1931 (H. J. Gibbons in **LTR**).

Juncus ambiguus Guss.
J. ranarius Song. & Perr.

Usually a coastal halophyte but also occasionally inland. It has been collected at Shapwick (8), 1933 (H. J. Gibbons in **LTR**), and Brewham (8), 1933 (Anon. in **K**) and at Berrow (9), 1945 (C & NS in **K**). The latter record was published as *J. bufonius* var. *congestus* Wahlb.

Juncus inflexus L. — Hard Rush

J. glaucus Sibth.

In wet places, particularly on basic soils. Very common throughout the county except on Exmoor in the extreme west.

Juncus effusus L. — Soft Rush

In wet places. Very common on acid soils, less so on basic ones but occurs throughout the county. The var. **compactus** Hoppe is frequent.

× **inflexus** (*J.* × *diffusus* Hoppe). A rare hybrid which has been recorded in North Somerset but has not been reliably identified during the recent survey.

Juncus conglomeratus L. Compact Rush

J. subuliflorus Drejer; *J. communis* E. Mey. pro parte

Damp heaths and marshes, particularly on acid soils. Frequent in all parts of the county.

× **inflexus.** This parentage has been suggested for plants collected near Taunton (3) by Dr W. Watson but there is considerable doubt as to whether this hybrid occurs in Britain.

Juncus maritimus Lam. Sea Rush

In salt marshes. Very rare and now only at Berrow (9). Formerly recorded also at Lilstock (2), in the Parrett estuary (3 & 5), at Portishead (9) and in the Avon estuary (10), but not seen in any of these places since 1920.

†**Juncus acutus** L. Sharp Rush

A plant of sandy shores for which there are very old records at Shurton Bars (2) (J. C. Collins) and Portishead (9) (S. Rootsey) which were doubted by both Murray and White as they were never confirmed by later botanists.

Juncus subnodulosus Schrank Blunt-flowered Rush

J. obtusiflorus Ehrh. ex Hoffm.

In peaty fen ditches. Locally common on West Sedgemoor (3), King's Sedgemoor (5), the peat moors (8), Kenn Moor and in the Gordano valley (9). Rare elsewhere and recorded recently only at Holme Moor (3), near Winscombe and Churchill (9), by the Chew Valley Lake (9) and near Shoscombe (10).

Juncus acutiflorus Ehrh. ex Hoffm. Sharp-flowered Rush

Marshes and wet meadows. Very common on moorland and other acid soils. Widespread but rare in the parts of the county where calcareous soils predominate.

× **articulatus** (*J.* × *surrejanus* Druce). A sterile hybrid known with certainty only on the peat moors (8) but probably more widespread where the distributions of the parent species overlap.

JUNCACEAE

Juncus articulatus L. Jointed Rush

J. lamprocarpus Ehrh. ex Hoffm.

In wet places. Very common throughout the county.

Juncus bulbosus L. Bulbous Rush

inc. *J. kochii* F. W. Schultz; *J. supinus* Moench

In bogs and on damp tracks on acid soils. Common on Exmoor and the other hilly moorlands of the west and south, on the peat moors (8) the eastern Greensand and the Mendip sandstone. Formerly recorded also in the Gordano valley (9) but not seen there recently. Recorded in all districts except 5.

Juncus subulatus Forsk.

A salt-marsh plant from the Mediterranean first recorded in Britain at Berrow (9) in May, 1957 by Dr A. J. Willis. There were two large patches when it was first found and these were subsequently identified on aerial photographs taken in April, 1954. As the area where it grows had not been long vegetated it was estimated it had only been established there for about ten years. The rush has since increased and is likely to persist so long as conditions in the marsh and the balance of salt and fresh water there remain unchanged.

Luzula DC.

Luzula pilosa (L.) Willd. Hairy Woodrush

L. vernalis DC.; *Juncoides pilosa* (L.) Kuntze

Woods and shady hedgebanks. Common over much of the county but rare in the lowlands. Recorded in all districts.

Luzula forsteri (Sm.) DC. Southern Woodrush

Juncoides forsteri (Sm.) Kuntze

In woods and on hedgebanks. Near Washford (2) and quite frequent on the red Keuper Marl of district 3. Unknown elsewhere in the county although there are rather doubtful old records from Wells (8) and Hutton (9).

× **pilosa** (*L.* × *borreri* Bromf. ex Bab.). This sterile hybrid has been recorded several times, but not recently, with the parent species in district 3.

Luzula sylvatica (Huds.) Gaudin — Great Woodrush

L. maxima (Reichard) DC.; *Juncoides sylvatica* (Huds.) Kuntze

Woodland and in damp rocky places on acid soil. Locally common in the combes of Exmoor and the Brendons (1, 2 & 3) it also occurs on the Greensand in districts 4, 6, 7 and 10, on the Mendip sandstone near Downhead (10) and on the northern sandstones between Clevedon (9) and Keynsham (10). Formerly recorded near Shipham (9) and Bath (10) but not seen in either locality for many years. Not recorded in district 5.

Luzula campestris (L.) DC. — Field Woodrush

Juncoides campestris (L.) Kuntze

Dry grassland. Very common throughout the county.

Luzula multiflora (Retz.) Lejeune — Heath Woodrush

L. erecta Desv.; *Juncoides multiflora* (Retz.) Druce

On moorland and in marshes on acid soils. Common in the west and south, on the peat moors (8), on the eastern Greensand and on the Mendip sandstone. Rare north of Mendip and seen recently only on the sandstone of Leigh Woods and Lord's Wood, Hunstrete (10). Babington's statement "not infrequent in damp places" in his *Flora Bathoniensis* (1834) is not true now even if it ever was. Not recorded in district 5.

AMARYLLIDACEAE

ALLIUM L.

Allium ampeloprasum L. — Wild Leek

Only on Steep Holm where its existence is now very precarious. It was referred to in an account book of the Manor of Norton Beauchamp written about 1625 as one of the dominant plants of the island which so tainted the rabbits they were not worth eating. It was still plentiful on one rocky slope in 1891 when the island was visited by Murray and White but is now reduced to only a few plants. An occurrence on the beach at Minehead Warren (2) in 1905 was probably due to seaborne or bird-sown seed; it persisted for some years but has now gone.

AMARYLLIDACEAE

Allium babingtonii Borrer Babington's Leek

A. ampeloprasum var. *babingtonii* (Borrer) Syme

Found by a ruined building on Porlock Marsh (2) in 1951 by Miss E. M. Medwin and still established there. An *Allium,* reported on the Marsh in 1916 by N. G. Hadden as *A. ampeloprasum* but eaten down by cattle before it could flower, may well have been the same.

Allium vineale L. Crow Garlic

Roadsides, waste places and a weed in arable and dry pasture fields. Common throughout the county except on the high moors. Mostly var. **compactum** (Thuill.) Cosson & Germ. with a head of bulbils only. Not recorded in district 1.

Allium oleraceum L. Field Garlic

Dry grassland. Rare and only reliably reported recently from a few spots in district 10 between Easton-in-Gordano and Bath. Formerly more widespread but recent reports from other parts of the county have not been supported by specimens except for one which proved to be the floriferous form of *A. vineale.*

Allium carinatum L. Keeled Garlic

An alien species planted on St Vincent's Rocks about 1897 by G. H. Wollaston which was first noticed on the Somerset side of the Avon Gorge (10) in 1962 by I. W. Evans and still persists there. Reported as a casual in a few other places.

Allium roseum L. Rosy Garlic

Garden escape.

Allium triquetrum L. Three-cornered Leek

A Mediterranean species long naturalized on roadside banks in Cornwall, and more recently in Devon, which was first recorded in the wild in Somerset near Minehead (2), 1958 (AD & OMH), and has since been reported in several places in that district. Also as a more obviously introduced plant near Bath (10).

Allium paradoxum (Bieb.) G. Don Few-flowered Leek

Another introduced plant established at Batheaston (10) since at least

1938 (AEW) which has since spread to several other places near Bath. Also established near Taunton (3) since 1952 (Miss Hector and Miss Roberts).

Allium ursinum L. Ramsons

Damp woods, shady hedgebanks and by streams. Locally abundant in woods in the north of the county and common elsewhere except in the extreme west and on the Levels.

LEUCOJUM L.

Leucojum vernum L. Spring Snowflake

In damp shady places. Very rare and only established in two places in Britain, one in Dorset and the other near Williton (2) where it was first noticed about 1910 by Miss M. A. Hellard and shown by her to Marshall in 1914. Probably originally introduced but now completely naturalized.

Leucojum aestivum L. Summer Snowflake

An uncommon garden escape in Somerset.

GALANTHUS L.

Galanthus nivalis L. Snowdrop

In damp woods and along the banks of streams where it often appears native although probably introduced originally, and on hedgebanks and roadsides where it is more obviously a garden escape. Frequent and recorded in all districts, but not recently in 1.

NARCISSUS L.

Narcissus pseudonarcissus L. Wild Daffodil

In woods and occasionally in fields. Rare and decreasing but still locally abundant where it survives. Recorded in all districts (not recently in 1) but more frequent on the Blackdown Hills (6) and in some of the Mendip woods (9 & 10) than elsewhere. Cultivated forms are quite common as escapes.

The introduced **N. poeticus** L. (Pheasant's Eye) and **N. × medioluteus** Mill. (*N. × biflorus* Curt.) (Primrose Peerless) were formerly much cultivated and became naturalized in several places but seem to have died out in all the old localities. The more modern *Narcissus* cultivars have been very seldom reported as escapes.

IRIDACEAE

SISYRINCHIUM L.

Sisyrinchium montanum E. L. Greene Blue-eyed Grass

S. angustifolium Mill. pro parte; *S. bermudiana* auct.

An alien naturalized on the golf course at Burnham and Berrow (9) where it was first recorded in 1906 (C. F. Vincent) and has since been reported at intervals until 1968 and may still persist. An occasional garden escape elsewhere.

IRIS L.

Iris foetidissima L. Stinking Iris

Woods, hedgebanks, limestone cliffs and dunes. Common on the Lias, less so on other calcareous soils. Recorded in all districts but not recently in 1.

Iris pseudacorus L. Yellow Flag

By rivers and streams, in marshes and damp meadows. Common throughout the county.

TRITONIA Ker-Gawl.

Tritonia × crocosmiflora (Lemoine) Nicholson Montbretia

Crocosmia × crocosmiflora (Lemoine) N. E. Br.

A widespread garden escape recorded in all districts except 7. It is particularly well established on the banks of the River Barle (1). It was first reported near Simonsbath in 1916 by Lady Davy and by 1918 it extended for nearly two miles (ESM). It now occurs at intervals over the whole length of the valley from Simonsbath to Dulverton.

GLADIOLUS L.

Gladiolus communis L. ssp. **byzantinus** (Mill.) A. P. Hamilton

G. byzantinus Mill.

Garden escape.

DIOSCOREACEAE

Tamus L.

Tamus communis L. Black Bryony

In hedges, wood borders and scrub. Very common throughout the county except on the higher parts of Exmoor.

ORCHIDACEAE

Cephalanthera Rich.

Cephalanthera damasonium (Mill.) Druce White Helleborine

C. pallens Rich.

In woods, usually under beech, on calcareous soils. Rare and decreasing. The only recent records are on the Lias near West Hatch (3) and on the Oolite round Bath (10) although there are quite a few older records in districts 5, 8 and 9 also. There is no obvious reason for the decline but this species is at the western limit of its British distribution in Somerset.

†**Cephalanthera rubra** (L.) Rich. Red Helleborine

Said to have been found by J. C. Collins on the Quantocks near Merridge (3) in 1836 but the record was never confirmed by any other botanist. There have been other reports of this very rare plant having been seen in the county but even if it once occurred its reappearance here is very unlikely.

Epipactis Zinn

Epipactis palustris (L.) Crantz Marsh Helleborine

In swamps where the water is alkaline and in dune slacks. Very rare and only recorded recently near Milverton and Pitminster (3) and at Berrow and near Winscombe (9). Formerly recorded in a number of other localities in districts 2, 4, 6, 8 and 10 but in none of these places had it a long history as it does where it still survives.

Epipactis helleborine (L.) Crantz Broad-leaved Helleborine

E. latifolia (L.) All.; *E. media* auct.

In woods and shady places. Quite frequent except in the lowlands. Recorded in all districts except 7 and not recently in 5.

ORCHIDACEAE

†**Epipactis purpurata** Sm. Violet Helleborine

E. sessilifolia Peterm.

In woods. Only once recorded in Somerset, near Quantock Lodge (3) in 1889 by H. S. Thompson. A specimen is in his herbarium at Birmingham University.

Epipactis leptochila (Godfery) Godfery Narrow-lipped Helleborine

In woods on calcareous soils. Very rare and only in the Cheddar Gorge (9). This species was only recognized as distinct in 1921 and by some authorities was then equated with *E. media* sensu Bab. but this was based on false premises. Plants named as "*E. media* Bab." were collected in the Gorge by G. C. Druce about 1882 and by Marshall in 1907, and on the strength of these records White claimed *E. leptochila* as a Cheddar plant. However, it was not till 1957 that indubitable *E. leptochila* was found there by Dr J. T. H. Knight. Since then two other colonies have been found some way from the first.

†**Epipactis phyllanthes** G. E. Sm. Green-flowered Helleborine

E. vectensis (T. & T. A. Stephenson) Brooke & Rose

Only once found in Somerset, on a roadside near Taunton (3) in 1891 by C. Bailey (specimen in **MANCH**). This species is considered by some botanists to be a variety of *E. leptochila*.

Spiranthes Rich.

Spiranthes spiralis (L.) Chevall. Autumn Lady's Tresses

S. autumnalis Rich.

In short grassland on dry calcareous slopes. Frequent on the Poldens and on the Mendips and other Carboniferous Limestone of north Somerset. Less so on other calcareous soils but recorded in all districts except 1 and 6. Rather erratic in appearance, and numbers vary considerably from year to year.

Listera R. Br.

Listera ovata (L.) R. Br. Twayblade

In woods and damp grassland. Rather common and recorded in all districts.

Listera cordata (L.) R. Br. Lesser Twayblade

On open moorland. Very rare and only on Exmoor, on Winsford Hill (1) and near the summit of Dunkery (1 or 2). Also recorded "on Brendon Hill near Chipstable" (3) in 1872 by Z. J. Edwards but not seen anywhere in the area since.

NEOTTIA Ludw.

Neottia nidus-avis (L.) Rich. Bird's-nest Orchid

A saprophyte of woods mainly on calcareous soils. Rather rare and not as frequent as it used to be perhaps because of clearance of old woodland. It has been recorded in all districts but not recently in 1, 5 or 7.

HERMINIUM Guett.

Herminium monorchis (L.) R. Br. Musk Orchid

Very rare and only seen recently on Bathford Hill (10) where it was found by Timothy Gaynor in 1971. The earliest Somerset record was on Odd Down, south of Bath (10) by Davis (Warner, 1802) and it was rediscovered in this neighbourhood by Miss A. E. White in 1938 but has not been seen there since 1946. The only other records were near Pilton (8) in 1892, and near Buckland Dinham (10) where it was last recorded in 1932.

COELOGLOSSUM Hartm.

Coeloglossum viride (L.) Hartm. Frog Orchid

Habenaria viridis (L.) R. Br.

In limestone grassland. Rare and apparently decreasing. It has only been seen recently at Hardington (4), White Sheet Downs (7) (Wilts), Shapwick (8), Charterhouse (9), Binegar and Wellow (10) though formerly recorded also in districts 1, 3, 5 and 6 as well as many other localities in districts 9 and 10.

GYMNADENIA R. Br.

Gymnadenia conopsea (L.) R. Br. Fragrant Orchid

Habenaria conopsea (L.) Benth.

In peaty fen meadows and on calcareous downland. Widespread but decreasing and now rather rare. Not recorded in district 1 nor recently in 7. Specimens of ssp. **densiflora** (Wahlenb.) G. Camus, Bergon & A. Camus, collected by Miss I. M. Roper in 1919 near Shapwick (8) and

Winscombe (9) are in her herbarium at Leeds University but this taxon has not been recorded since although the type has been seen recently in both localities.

PLATANTHERA Rich.

Platanthera chlorantha (Custer) Reichb. Greater Butterfly Orchid

Habenaria chloroleuca Ridl.; *H. virescens* Druce

In woods and occasionally in grassland and on road verges. Frequent and widespread on calcareous soils. Not recorded recently in district 1.

Platanthera bifolia (L.) Rich. Lesser Butterfly Orchid

Habenaria bifolia (L.) R. Br.

On heaths and in boggy places usually on acid soils but occasionally on calcareous ones. On the Blackdown Hills (3, 4 & 6) and the peat moors (8). Rare elsewhere and only recorded recently in a few localities in districts 1 and 7. There are older records also in districts 5 and 9 but some of these are suspect because of confusion with *P. chlorantha.*

OPHRYS L.

Ophrys apifera Huds. Bee Orchid

Limestone grassland, old quarries and stabilized dunes. Frequent in the eastern half of the county but rather rare in the west and more or less confined there to the coastal Lias in district 2. Not recorded in district 1 nor recently in 6.

× **insectifera** (*O.* × *pietzschii* Kümpel). A plant found in an old quarry in Leigh Woods (10) in 1968 by M. Flower, I. D. R. Stevens and M. C. Whiting is the only example of this hybrid known at present to have occurred naturally in the wild. The record was first published as an unusual form of *O. insectifera* but hybridity was suspected. The plant was seen subsequently by several other botanists but it was not till 1976 that its identity was confirmed by comparison with a hybrid produced by artificial cross-pollination in the wild in East Germany and first described by H. Kümpel in 1971.

Ophrys insectifera L. Fly Orchid

O. muscifera Huds.

Wood borders and grassland on limestone. Rare and only near Blue

Anchor (2), Crewkerne (4), Tickenham (9), Leigh Woods and around and to the south of Bath (10). In all these areas it has been known for a very long time.

HIMANTOGLOSSUM Koch

Himantoglossum hircinum (L.) Spreng. Lizard Orchid

Orchis hircina (L.) Crantz

At the western extreme of its British distribution in Somerset. It was first recorded in the county in the grounds of a derelict building at Burnham (8) in 1923 (Miss Stewart) and persisted there till at least 1951. Otherwise its appearances in the county have been sporadic, at Charlton Mackrell (5) in 1976, at Penselwood (7) in 1956 and at Tickenham (9) in 1978. Two of these occurrences were in gardens but the plant was not deliberately introduced. There have been one or two other unverified reports.

ORCHIS L.

†**Orchis ustulata** L. Burnt Orchid

In limestone grassland. Probably now extinct. Only recorded in three localities in the county during the last hundred years, on Wavering Down (9), 1892 (W. F. Miller), on Callow Hill (9), 1959 (John Hodgson), and on Claverton Down near Bath (10), ca. 1920 (NYS). It was apparently more widespread formerly in district 9 and had been known near Bath since about 1760.

Orchis morio L. Green-winged Orchid

In grassland mainly on calcareous soils. Widespread and recorded in all districts except 1 and 6, it is certainly not common now although classed as such by Murray (1896). This species has decreased nationally due to the ploughing of old grassland.

Orchis mascula (L.) L. Early Purple Orchid

In woods and on hedgebanks. Common over much of the county where there are suitable habitats.

DACTYLORHIZA Necker ex Nevski

Dactylorhiza fuchsii (Druce) Soó Common Spotted Orchid

ORCHIDACEAE

Dactylorchis fuchsii (Druce) Vermeul.; *Orchis fuchsii* Druce; *O. maculata* auct.

In woods and grasslands. Common on calcareous soils, less so on neutral and slightly acid ones but widespread and recorded in all districts.

× **maculata** (*D.* × *transiens* (Druce) Soó). A rare hybrid which may occur where the parent species grow together which they seldom do in Somerset due to their different ecological requirements. A colony of the two species with some intermediates was seen near Chelwood (10) in 1966. Previous records dated 1926 and 1935 were from the Blackdown Hills (4 & 6).

× **majalis** (*D.* × *grandis* (Druce) P. F. Hunt). A fairly frequent hybrid in marshes and wet meadows where the parent species meet, which has been recorded in all districts except 1 and 7.

Dactylorhiza maculata (L.) Soó Heath Spotted Orchid

Dactylorchis maculata (L.) Vermeul ssp. *ericetorum* (E. F. Linton) Vermeul; *Orchis ericetorum* (E. F. Linton) E. S. Marshall; *O. maculata* auct.

Moors, heaths and meadows on acid soils. Common on the moors and hills of the west and south, on the peat moors (8), the eastern Greensand and on the Mendip sandstone. Rare elsewhere but it occurs on limestone heath where locally acid conditions have developed due to leaching. Sometimes reported in error for *D. fuchsii;* the separation of the two species of Spotted Orchid was not generally accepted by British botanists until comparatively recently.

× **majalis** (*D.* × *hallii* (Druce) Soó). This hybrid might be expected where the parent species occur together but has not been reported in the county recently. It was recorded by Marshall near Stogumber (2) in 1910, and at Britty Common (4) in 1926 and 1935 (WW).

Dactylorhiza incarnata (L.) Soó Early Marsh Orchid

Dactylorchis incarnata (L.) Vermeul; *Orchis incarnata* L.; *O. strictifolia* Opiz

In bogs and wet fen meadows. Rare. The distribution is imperfectly known because of confusion with *D. majalis* but it certainly grows on the peat moors (8) and in dune slacks at Berrow (9). There are old records in districts 1, 3, 4 and 6 also.

× **majalis** (*D.* × *wintoni* (A. Camus) P. F. Hunt). This hybrid was recorded on Shapwick Heath (8) where the parent species both grow in 1947 (C & NS).

Dactylorhiza majalis (Reichb.) Hunt & Summerh. ssp. **praetermissa** (Druce) D. Moresby Moore & Soó Southern Marsh Orchid

D. praetermissa (Druce) Soó; *Dactylorchis praetermissa* (Druce) Vermeul; *Orchis praetermissa* Druce; *O. latifolia* auct.

In marshes and damp meadows. Widespread and frequent but decreasing, no doubt due to improved drainage. Recorded in all districts.

DACTYLORHIZA × PLATANTHERA

An orchid found near Pilton (8) in 1913 was believed to be a hybrid and named as × **Rhizanthera somersetensis** (Camus) Soó, with putative parentage *Dactylorhiza fuchsii* × *Platanthera bifolia.* Further study of the preserved specimen suggests that it is in fact a variant of *D. fuchsii.* If it were a hybrid a more probable parentage would be *D. fuchsii* × *P. chlorantha* as *P. bifolia* does not occur in the area. Neither combination has been known to have occurred anywhere else.

ACERAS R. Br.

†**Aceras anthropophorum** (L.) Ait. f. Man Orchid

On limestone downs. Probably extinct in Somerset. Found near South Stoke (10) in 1933 by Miss A. E. White who did not reveal the exact locality to anyone else and it has not been seen since. The specimen is in the British Museum herbarium. The only other record was between Portishead and Clevedon (9), ca. 1892, when unfortunately all the plants were dug up. The descendants of these were seen by White in a garden in Clifton, Bristol, in 1907, but none were ever seen in the wild again.

ANACAMPTIS Rich.

Anacamptis pyramidalis (L.) Rich. Pyramidal Orchid

Orchis pyramidalis L.

Calcareous grassland. Common on the Lias and Oolite, frequent on the Carboniferous Limestone and occasional elsewhere. Recorded in all districts except 1 and not recently in 7.

ARACEAE

ACORUS L.

Acorus calamus L. Sweet Flag

By the sides of rivers and ponds. Originally introduced but now well established by the Avon from Bath to below Keynsham (10), it was first recorded there by Davis (Warner, 1802). It is also established by a few ornamental ponds in districts 9 and 10. It was recorded as plentiful on King's Sedgemoor (5) by J. C. Collins and on the peat moors (8) by Sole (Collinson, 1791) but has not been seen in either place by later botanists.

CALLA L.

Calla palustris L. Bog Arum

A garden escape naturalized on a damp roadside near South Stoke (10).

ARUM L.

Arum maculatum L. Lords and Ladies

Woods, hedgebanks and shady places. Very common throughout the county except on the higher parts of Exmoor.

Arum italicum Mill. Italian Lords and Ladies

Planted in private grounds and sometimes seen as a throwout from such places.

DRACUNCULUS Mill.

Dracunculus vulgaris Schott Dragon Arum

An evil-smelling plant from the Mediterranean which is sometimes grown in gardens for its striking appearance. A plant appeared in the newly-made small garden of a new house in Bath (10) in 1967 (Miss E. H. Stevenson) probably brought in with topsoil from somewhere else.

LEMNACEAE

SPIRODELA Schleid.

Spirodela polyrhiza (L.) Schleid. Greater Duckweed

Lemna polyrhiza L.

In stagnant ditches, rhines, ponds and canals. Common throughout the

Levels, in the Minehead marshes (2) and in the Kennet & Avon Canal near Bath (10). Apart from the latter at about 100 ft., there are very few records above the 50-foot contour and these are usually casual occurrences in isolated ponds. This species had never been known to flower in Britain until in July, 1906, Mrs. Gregory and Miss Peck collected flowering fronds near Wedmore (8). There have been no similar reports recently. Not recorded in districts 1, 6 or 7.

LEMNA L.

Lemna trisulca L. Ivy-leaved Duckweed

In similar places to *Spirodela polyrhiza*. The distribution is very much the same but it is more common and it is not quite so strictly limited to the Levels and other places at low altitudes.

Lemna minor L. Common Duckweed

In ditches, ponds and other stagnant water. Very common throughout most of the county but less so on Exmoor.

Lemna gibba L. Fat Duckweed

In similar places to the other duckweeds. The least plentiful of the *Lemna* species but common enough on the Levels (3, 4, 5, 8 & 9). Away from the main marshlands it has only been recorded recently at Porlock Marsh (2), Barrow Gurney and at Claverton (10) in the Kennet & Avon Canal.

WOLFFIA Hork. ex Schleid.

Wolffia arrhiza (L.) Hork. ex Wimm. Rootless Duckweed

On still water, often with other duckweeds. First found in Somerset near Taunton (3) in 1911 by W. Watson, it was reported in 1915 on the peat moors (8) and in other places on the Levels south of Mendip (9). In 1921 it was seen near Kenn (9) and in 1929 near Thorney (4) on the southern moors. By 1952 Mrs C. I. & N. Y. Sandwith wrote that it was frequent and well distributed on the North Somerset Levels. It is still rare and local in South Somerset and was not recorded in district 5 until 1969. It is now more plentiful in the area round Kingston Seymour (9) than anywhere else. It persisted in the original locality for many years but disappeared some time after 1944.

SPARGANIACEAE

SPARGANIUM L.

Sparganium erectum L. Branched Bur-reed
S. ramosum Huds.

In and by ditches, rivers, streams and ponds. Very common over most of the county but rather sparse in the far west. The species has been divided into a number of subspecies, mainly based on the fruit shape, but the distribution of these has not been worked out; ssp. **neglectum** (Beeby) Schinz & Thell. and ssp. **microcarpum** (Neuman) Domin have both been identified as occurring in the county.

Sparganium emersum Rehm. Unbranched Bur-reed
S. simplex Huds. pro parte

In ditches, streams and rivers, often in deeper and more swiftly-flowing water than *S. erectum*. Common except in the west. Not recorded in districts 1 or 6 nor recently in 2.

Sparganium minimum Wallr. Least Bur-reed

In peaty pools. Very rare and only on the peat moors (8) where it is extremely scarce and local.

TYPHACEAE

TYPHA L.

Typha latifolia L. Reed Mace or False Bulrush

In ditches, old peat-diggings, ponds and swamps. Common over most of the county but thins out in the west.

Typha angustifolia L. Lesser Reed Mace

Ditches and ponds. Very rare and only seen recently at Wembdon and North Moor, North Petherton (3), Cricket St Thomas (6) and Berrow (9). The species was never seen in Somerset by either Murray or White but there are former records in districts 2, 4, 5, 8 and 10. Narrow-leaved forms of *T. latifolia* are sometimes reported as this in error.

CYPERACEAE

ERIOPHORUM L.

Eriophorum angustifolium Honck. Common Cottongrass

E. polystachion L. nom. ambig.

In bogs and on acid heathland. Still common on Exmoor and the Brendon, Quantock and Blackdown Hills, on the peat moors (8) and on the Mendip sandstone. Rare and decreasing elsewhere. Not recorded in district 5 nor recently in 7.

†**Eriophorum gracile** Koch ex Roth Slender Cottongrass

Found in a very wet swamp on Britty Common (4) in 1915 by Marshall, growing quite close to the other three species of Cottongrass. It was last seen there in 1924 since when the area has been considerably changed by drainage and afforestation.

Eriophorum latifolium Hoppe Broad-leaved Cottongrass

In fen swamps. Very rare and only seen recently near Milverton (3) and Winscombe (9) though there are older records in districts 1, 4, 6 and 10.

Eriophorum vaginatum L. Hare's-tail Cottongrass

In acid bogs. Common on Exmoor (1 & 2). It also occurs in small quantity on the Blackdown Hills (4 & 6), the peat moors (8) and the Mendip sandstone (9 & 10) but in all these areas there is less than there used to be. Old records which have not been confirmed recently were at Durleigh (3), Witham (8) and Weston-in-Gordano (9).

SCIRPUS L.

Scirpus cespitosus L. Deergrass

Trichophorum cespitosum (L.) Hartm.

On acid moors, particularly on the higher ground. Common on Exmoor (1 & 2), Haddon Hill (1) and on the Blackdown Hills (3 & 6). Rare elsewhere and recorded recently only at the north end of Quantock (2), near Priddy and on Blackdown on Mendip (9), and on Highridge Common (10) near Bristol. Formerly in a small area of Westhay Moor (8) but exterminated by peat cutting about 1972.

CYPERACEAE

Scirpus maritimus L. Sea Club-rush

In ditches, salt marshes, by streams, rivers and ponds. Common in brackish ditches and on the inner margins of salt marshes along the coast and up the tidal estuaries from Porlock (2) to Portbury and Ashton Gate (10). Remarkable in Somerset for its plentiful occurrence inland on the Levels as far as Kingsbury Episcopi (4), Ilchester (5) and Glastonbury (8). It is similarly found inland in Dorset in the Stour valley and it occurs on the Somerset side of the border near Templecombe (7). Also in the Blagdon (9) and Chew Valley (10) reservoirs. There are a few isolated records by artificial ponds in districts 4 and 5 away from the Levels where it was presumably introduced. A record in the Avon near Bath by T. B. Flower in Babington (1839) was probably a casual occurrence as there have been no later records in the non-tidal part of the river.

Scirpus sylvaticus L. Wood Club-rush

In marshy woods and fields and by streams. Rather rare and decreasing. It is still to be found in wilder and less cultivated places near Dulverton (1), Wellington (3), on the Blackdown Hills (4 & 6), on the eastern borders from Wincanton (7) to Berkley (10) and in a few spots near Bath. Not recorded in districts 5 and 9, nor recently in 2.

Scirpus holoschoenus L. Round-headed Club-rush

Holoschoenus vulgaris Link

Sandy places by the sea. Very rare and now only in one locality. Ray stated in his *Historia* (1688) that it had recently been found in Somerset by D. Stephens but in a later work this was corrected to Braunton Burrows in Devonshire where the plant still grows. Sole in 1791 wrote that it grew "near the seaside below Watchet" (2) and there are specimens extant from this locality dated as late as 1849 and Marshall had an unlocalized Somerset one dated 1862. It was believed to be extinct in the county until 1896 when Mrs E. S. Gregory found a single clump on the dunes at Berrow (9). In spite of threats and rumours of its extinction this clump still survives. In 1927 an attempt was made to sow seed from Braunton in its vicinity but no seedlings appeared.

Scirpus lacustris L.

ssp. **lacustris** Common Club-rush, Bulrush

Schoenoplectus lacustris (L.) Palla

In rivers and lakes. Local and only in the Parrett, Isle and Yeo (4 & 5), Cale (7), Brue (8), Avon and Frome (10). Also in the Blagdon (9) and Chew Valley (10) reservoirs and in a pond at Upton Noble (10). There are a few doubtful records elsewhere which are probably errors for ssp. *tabernaemontani.* Formerly recorded at Ford Gate (3), 1857 (Hb. T. Clark) but not seen there since.

ssp. **tabernaemontani** (C. C. Gmel.) Syme Grey Club-rush

S. tabernaemontani C. C. Gmel.; *Schoenoplectus tabernaemontani* (C. C. Gmel.) Palla

In ditches near the sea and in moor ditches on the Levels. Rather rare and local. At intervals near the coast; Porlock and Minehead (2), Pawlett Hams (5), Berrow, Bleadon and Wick St Lawrence (9). Inland it is plentiful on West Sedge Moor (3) and on King's Sedge Moor (5) particularly at its eastern end. It is also on Congresbury, Kenn, Nailsea and Tickenham Moors (9). It has not been seen on Cheddar Moor (9) since 1901, nor in the Gordano valley since 1951. Marshall recorded it at Chard Reservoir (4) but it has gone from there.

Scirpus setaceus L. Bristle Club-rush

Isolepis setacea (L.) R. Br.

Damp muddy places on sandy or gravelly soils. Quite frequent on acid moorland. Rather rare elsewhere but recorded in all districts except 5.

Scirpus cernuus Vahl Slender Club-rush

S. numidianus sensu Murray; *Isolepis cernua* (Vahl) Roem. & Schult.

In muddy places near the sea. Very rare and only seen recently near Walton- and Weston-in-Gordano (9). Formerly in several spots near Minehead (2) but not seen there since 1933, and at Berrow (9) where it was last recorded in 1945.

Scirpus fluitans L. Floating Club-rush

Eleogiton fluitans (L.) Link

In ditches and streams on acid soils. Very rare and only seen recently on the peat moors (8) and in the Gordano valley (9). There are old records from Exmoor (1), the Quantocks (2 & 3) and Blackdown Hills (4 or 6), North Newton (3) and Blackdown on Mendip (9) but it has not been seen in any of these places since 1918.

CYPERACEAE

ELEOCHARIS R. Br.

†Eleocharis acicularis (L.) Roem. & Schult. Needle Spike-rush

Sides of canals and ponds. Always very rare and now probably extinct. Known for many years on the edge of the canal between Bath and Claverton (10) but not recorded there since 1945. It was once seen in the Bridgwater & Taunton Canal near Maunsel (3) in 1907.

Eleocharis quinqueflora (F. X. Hartmann) Schwartz
Few-flowered Spike-rush

E. pauciflora (Lightf.) Link; *Scirpus pauciflorus* Lightf.

In bogs and marshes. Very rare and only seen recently near Chetsford Water (2), 1973 (Mrs E. Hesselgreaves det. A. C. Jermy) and at Berrow where it has been known since 1897. There are a number of older records thinly scattered in districts 1, 3, 4, 6, 8 and 9.

Eleocharis multicaulis (Sm.) Desv. Many-stalked Spike-rush

In wet peaty bogs. Frequent on Exmoor (1 & 2), the northern end of Quantock (2 & 3) and the peat moors (8). Also recorded on the Blackdown Hills (3) and Chard Common (6). Not seen recently on Mendip (9 & 10).

Eleocharis palustris (L.) Roem. & Schult.
ssp. **vulgaris** Walters Common Spike-rush

In marshy places. Common throughout the county.

Eleocharis uniglumis (L.) Schult. Slender Spike-rush

In brackish marshes near the sea and in a calcareous marsh inland. Rare and recorded recently only near Shepton Mallet (8), Lympsham, Wick St Lawrence and in the Gordano valley (9). First recorded in the county at Berrow (9) in 1894 (S. T. Dunn) but not noted there since 1896. Other former records were on Tickenham and Nailsea Moors (9) in 1947 and 1950 respectively (C & NS).

BLYSMUS Panz.

Blysmus compressus (L.) Panz. ex Link Flat-headed Sedge

Scirpus caricis Retz.

In marshes on calcareous soil. Very rare and now only known in Somerset in one locality, near Shepton Mallet (8). Formerly recorded also at Burnham (8), Weston-super-Mare (9), Leigh Woods and at Bathampton and Batheaston (10) but now gone from all these places.

CYPERUS L.

Cyperus longus L. Galingale

In ponds and marshes. Believed to have been native near Walton-in-Gordano (9) where it was first recorded by Sole in 1782 in what was then a pond. The place was drained and planted with potatoes in 1882 but the plant did not finally disappear till after 1887. Sometimes planted in ornamental ponds and occasionally found as an escape.

Cyperus fuscus L. Brown Galingale

By muddy ditches and ponds. Very rare and only in the Gordano valley (9) where it was first discovered in 1900 by S. J. Coley. It varies very much in quantity from year to year.

Cyperus eragrostis Lam.

Garden escape.

SCHOENUS L.

Schoenus nigricans L. Black Bog-rush

In boggy places where the water is alkaline. Very rare and now only near Winscombe and Churchill (9). Also recorded near Bathealton (3), on Shapwick Moor (8) and on the shore between Clevedon and Portishead (9) but not seen recently in any of these places.

RHYNCHOSPORA Vahl

Rhynchospora alba (L.) Vahl White Beak-sedge

In acid bogs. Rare and local on the Blackdown Hills (3, 4 & 6) and the peat moors (8). Formerly also on the Mendip sandstone (9) but not recorded there since 1927.

†**Rhynchospora fusca** (L.) Ait. f. Brown Beak-sedge

In acid bogs. Probably extinct. It was recorded on the peat moors (8) by Sole in 1782 but soon became very rare though it was seen from time to time between 1832 and 1931. It is believed there were two localities, on Shapwick Heath and Westhay Moor respectively. It was rediscovered on the latter by Mrs C. I. and N. Y. Sandwith in 1947. The enclosure in which it grew was leased by the Somerset Trust for Nature Conservation for some years but the lease lapsed in 1969 and the plant was destroyed by peat cutting soon afterwards. There is still a faint hope that it may survive in the Shapwick locality but it has not been seen there since 1931 despite many searches.

CLADIUM Browne

Cladium mariscus (L.) Pohl — Sedge

In fens. Once plentiful on the Levels and the plant which gave Sedgemoor its name. Sole (1791) said it was then abundant on King's Sedge Moor (5) and (1782) that it grew between Burtle, Wedmore and Glastonbury (8), but by 1896 Murray believed it to be extinct due to the drainage of the Levels. It was found again on Butleigh Moor (5) in 1955 by Mrs D. H. Perrett and E. J. Hamlin and it still survives there in small quantity. Similarly it was refound near Burtle (8) in 1910 by H. Corder, and later by others in several other places on the peat moors in at least two of which it still exists. It also grows in a fen marsh near Wiveliscombe (3) where it was discovered by Marshall in 1910.

CAREX L.

Carex laevigata Sm. — Smooth-stalked Sedge

In marshy fields and wet wooded valleys on acid soils. Frequent in Exmoor and Brendon combes (1, 2 & 3), on the Blackdown Hills (3, 4 & 6) and on the eastern Greensand (7, 8 & 10). Very rare elsewhere. Some reports from outside its main area of distribution have proved to be *C. distans* or *C. binervis.*

Carex distans L. — Distant Sedge

In marshy places near the sea and by alkaline springs inland. Frequent near the coast between Stolford (2) and Portishead (9). Rather rare in the lowlands further inland in a few places in districts 3, 5, 8, 9 and 10.

Carex hostiana DC. — Tawny Sedge

C. hornschuchiana Hoppe; *C. fulva* auct.

In peaty bogs and marshes with base-rich water. Locally frequent in the

Exe valley (1), near Wiveliscombe (3), on the Blackdown Hills (3, 4 & 6), on the eastern borders (7, 8 & 10) and on the peat moors (8). It also occurs at a few places on Mendip (9) and around Bath (10). Formerly recorded in district 5 but not seen there recently.

Carex binervis Sm. Green-ribbed Sedge

On acid moors. Common on Exmoor, the Brendon, Quantock and Blackdown Hills, on the eastern Greensand and the Mendip sandstone. Elsewhere *C. distans* has been mistaken for this species. It formerly occurred with other acid-moor species on some of the hilltops near Bath (10) but has not been seen there for over a century.

Carex lepidocarpa Tausch Long-stalked Yellow Sedge

Base-rich meadows and flushes. Rare and local. Near Milverton (3), 1968 (RGR), Shepton Mallet (8) and Winscombe (9). Rather more frequent in district 10 mainly on the Oolite. The records cited by Murray (1896) and White (1912) as *C. flava* type or var. *elatior* belong here; the true *C. flava* L. does not occur in Somerset.

Carex demissa Hornem. Common Yellow Sedge

C. flava var. *minor* Towns.

On boggy moorland, in marshy places and by springs and streams on acid soils. Common on the moors in the west and south of the county, on the peat moors, on the eastern Greensand and the Mendip sandstone. Rare elsewhere though it does occur in a few spots on the sandstone in the north. Not recorded in district 5.

× **hostiana.** A highly sterile hybrid, frequent in Britain where the parent species grow together. There are several records for it in Somerset but it has not been identified in the county recently.

× **serotina.** On the peat moors (8) where in 1951 Mrs C. I. and N. Y. Sandwith reported it as more plentiful than *C. serotina.*

Carex serotina Mérat Small-fruited Yellow Sedge

C. oederi auct.

In peaty fens. Very rare and only on the peat moors (8).

CYPERACEAE

Carex extensa Gooden. Long-bracted Sedge

Brackish marshes near the sea. Rather rare and only in district 9; Berrow, Kingston Seymour, Clevedon and Portishead.

Carex sylvatica Huds. Wood Sedge

Woodland. Very common throughout the county except for parts of Exmoor and of the Levels where there is no suitable habitat.

Carex depauperata Curt. ex With. Starved Wood Sedge

Woods and hedgebanks on limestone. Very rare. One or two clumps existing precariously in a lane near Cheddar (9), where it has been known since 1860, are the only surviving representatives of the species in the wild in Great Britain. The plant was found in Ireland in 1973. Old records from near Templecombe (7) and in Leigh Woods (10) have never been confirmed.

Carex pseudocyperus L. Cyperus Sedge

In fen ditches and pools. Common on the Levels of districts 3, 4, 5, 8 and 9, especially where peat is exposed in the ditches. Occasionally by pools and ponds away from the main area of distribution but seldom long persistent in such places.

Carex rostrata Stokes Bottle Sedge

In and fringing pools of acid water and in bogs. Frequent on Exmoor (1 & 2) and the peat moors (8). Rare elsewhere and only in a few spots in the Gordano valley (9) and on the Mendip sandstone (9 & 10).

Carex vesicaria L. Bladder Sedge

In drainage ditches on the Levels. Very rare. Near North Curry (3), 1974 (JGK), Kingsbury Episcopi and Langport (4), Axbridge (9) and Witham Friary (10), 1969 (CAH). Not known in the county until discovered by Dr H. Downes near Langport (4) in 1924. In 1946 when W. B. Waterfall's herbarium was presented to Kew a specimen from Axbridge Moor dated 27th June, 1877, was found in it and Mrs C. I. and N. Y. Sandwith relocated it there (where it still survives) in 1947.

Carex riparia Curt. Greater Pond Sedge

In marsh ditches and by streams and rivers. Very common throughout the

Levels and frequent elsewhere except west of Taunton (3) where there are very few records. Not recorded in district 1.

Carex acutiformis Ehrh. Lesser Pond Sedge

C. paludosa Gooden.

In marshy fields, by ditches, streams and pools. Common over much of the county except in the west (1 & 2) where it is rare.

Carex pendula Huds. Pendulous Sedge

Damp woods, roadside ditches and shady hedgebanks. Common throughout the county except on the highest ground and over much of the Levels.

Carex strigosa Huds. Thin-spiked Wood Sedge

Damp woods, muddy tracks and by streams, on calcareous clay. Widespread but not common. Absent from the west (1 & 2) and the Levels.

Carex pallescens L. Pale Sedge

Damp pastures and open woodland on clay. Rather rare except on the Clay-with-Flints of the Blackdown Hills (3, 4 & 6) and on the Oxford Clay east of Bruton (7, 8 & 10). Not recorded in district 5 nor recently in 2.

Carex panicea L. Carnation Sedge

In bogs and marshes. Common on acid soils. Less so on calcareous ones but widespread and recorded in all districts.

†**Carex limosa** L. Bog Sedge

A record by Swayne from a peat bog on the Mendip Hills, quoted by Withering (1796) and the *Botanist's Guide* (1805) has never been confirmed by anyone else and may have been an error.

Carex flacca Schreb. Glaucous Sedge

C. glauca Scop.

In grassland. Very common on limestone downs, less frequent elsewhere but occurs in a wide range of habitats throughout the county.

CYPERACEAE

Carex hirta L. Hairy Sedge

In damp grassland, pastures and on road verges. Common throughout much of the county but less frequent in the west.

Carex lasiocarpa Ehrh. Slender Sedge

C. filiformis auct.

Peaty fen swamps. Very rare and only on the peat moors (8) where it was first collected by T. Clark in 1855.

× **riparia** (*C.* × *evoluta* Hartm.). Discovered on the peat moors at Sharpham (8) in 1915 by H. S. Thompson. It was still plentiful there in 1955 but its survival has not been confirmed since. This is the only known occurrence of this hybrid in the British Isles.

Carex pilulifera L. Pill Sedge

On moorland with peaty or leached soils. Common on Exmoor (1 & 2) and frequent on the Quantock and Blackdown Hills (2, 3, 4 & 6) and on the eastern Greensand (7, 8 & 10). Rare on the peat moors (8), on the Mendip and Pennant Sandstone (9 & 10) and on Broadfield Down (9).

Carex caryophyllea Latourr. Spring Sedge

C. verna Chaix; *C. praecox* auct.

Mainly in dry grassland but occasionally in damper places. Very common on the calcareous downs of North Somerset. Less frequent elsewhere but widespread and recorded in all districts.

Carex montana L. Soft-leaved Sedge

Rough grassy places on limestone. This rare sedge is quite plentiful over a small area of Mendip near Charterhouse (9) where it was first found by the Rev E. F. Linton in 1890. White quotes Sole's record (Collinson, 1791) of *"C. montana"* for rocks opposite Hotwells (10) and concludes "some mistake was made" but Sole probably followed Hudson's *Flora Anglica* where the name is applied to *C. pilulifera*.

Carex humilis Leyss. Dwarf Sedge

On rocky limestone downs. Rare and local. Only on Brean Down, where it is locally plentiful, and Crook Peak (9) and in the Avon Gorge (10). It was also recorded on Bathampton Down (10) by Miss Peck and vouched for by White himself in his *Bristol Flora* (1912) but although searched for many times by Bath botanists it has never been refound.

Carex digitata L. Fingered Sedge

In woods on limestone. Only in the Avon Gorge (10) where it has been known since 1799 and from whence many specimens have been collected but where it still survives, fortunately in rather inaccessible places. An older record by Sole near Bath, at Friary Wood, Hinton Charterhouse (10) is referred to by Hudson (1778) and quoted by later authors but it has not been seen there since 1838. The wood has been largely replanted since then and although some of the species recorded by Sole are still there in the least disturbed parts there is little hope of this one being refound.

Carex elata All. Tufted Sedge

C. stricta Gooden.; *C. hudsonii* A. Benn.

Fen ditches. Very rare and only seen recently near Axbridge (9) where it was first collected by W. B. Waterfall in 1877 but named by him as *C. paludosa* Gooden. (*C. acutiformis* Ehrh.). It was correctly identified when his herbarium was presented to Kew in 1946, and Mrs C. I. and N. Y. Sandwith rediscovered the plant in 1948. It formerly grew near Tickenham and in the Gordano valley (9) but was last reported in these places in 1938 and 1956 respectively. Very old records from near Bath (10) have never been confirmed.

Carex acuta L. Slender Tufted Sedge

C. gracilis Curt.

In moor ditches, by ponds and rivers. Rare and recently recorded only at Long Sutton (5), Berrow, Axbridge (9) and Witham Friary (10). It formerly grew by the Chew at Compton Dando and by the Avon between Warleigh and Keynsham (10) but has not been recorded in this area since 1945.

× **nigra.** A specimen collected by C. A. Howe at Witham Friary (10) in 1975 was identified as this hybrid by A. C. Jermy.

Carex nigra (L.) Reichard Common Sedge

C. goodenowii Gay

In marshes, ditches and bogs. Common on acid soils. Less frequent but widespread on other soils where locally acid conditions have developed in poorly-drained places or stagnant ditches.

Carex paniculata L. Greater Tussock Sedge

In woodland swamps, by ponds and in very wet marshes. Common on the Blackdown Hills (3, 4 & 6) and on the peat moors (8). Rare elsewhere but there are a few scattered records in all districts except 5.

Carex diandra Schrank Lesser Tussock Sedge

C. teretiuscula Gooden.

Peaty meadows. Very rare and only on the peat moors (8) where it is decreasing and its survival is threatened. Old records from near Weston-super-Mare (9) and Bath (10) were almost certainly errors.

Carex otrubae Podp. False Fox Sedge

C. vulpina auct.

By ditches and in marshy places. Very common in the lowlands and frequent elsewhere except in the west. Not recorded in district 1.

× **spicata** (*C.* × *haussknechtii* Senay). A plant collected near Whitchurch (10) in 1925 by H. S. Thompson was believed to be this hybrid but there is some doubt as to whether it really is.

× **remota** (*C.* × *pseudoaxillaris* K. Richt.; *C.* × *axillaris* auct.) This hybrid has been recorded in moor and roadside ditches in all districts except 1, 2 and 4 but it is not common.

Carex disticha Huds. Brown Sedge

Marshy fields. Frequent on the Levels in districts 3, 5, 8 and 9 and in the Cale basin near Wincanton (7). Rare elsewhere and not recorded in districts 1, 2 or 6, nor anywhere in the county west of Taunton (3).

Carex arenaria L. Sand Sedge

On sandy shores. Minehead, Dunster and Steart (2) and very common on the dunes all along the coast between Burnham (8) and Sand Bay (9). Once recorded as a casual by the railway below the Clifton suspension bridge (10), 1955 (IWE).

Carex divisa Huds. Divided Sedge

In marshes near the sea. Very rare and now only at Berrow (9). Formerly at Minehead Warren (2) (1891), Shapwick Heath (8) (1906), Kewstoke Bay (9) (1894) and Ashton Gate (10) (1933); the dates given are those of its most recent records in these places.

Carex divulsa Stokes Grey Sedge

Hedgebanks and roadsides. Common throughout the county except in district 1 where it is rare.

ssp. **leersii** (Kneucker) W. Koch

C. polyphylla Kar. & Kit.; *C. muricata* ssp. *leersii* Aschers. & Graebn.

In similar places to the type. The existence of this taxon in North Somerset was established in 1946 when E. Nelmes confirmed specimens collected near Tickenham (9), 1920 (C & NS), and Axbridge (9), 1936, as this. It has been seen recently in both spots. He also identified a plant from Batheaston (10), 1958 (G. W. Garlick). It has since been found also near Wellow (10), 1979 (DG conf. R. W. David).

(Marshall identified a plant he found near West Monkton (3) in 1914 as the hybrid *C. divulsa* × *C. spicata* but this has never been confirmed and the hybrid has not been found anywhere else.)

Carex spicata Huds. Spiked Sedge

C. contigua Hoppe; *C. muricata* auct.

On hedgebanks and in rough grassland on calcareous soils. Common in the eastern half of the county but rather rare west of Taunton (3). Recorded in all districts.

Carex muricata L. ssp. **lamprocarpa** Čelak. Prickly Sedge

C. pairaei F. W. Schultz

The distribution of this species is uncertain because of difficulties of identification and confusion with *C. spicata* from which it was not separated by either Murray or White. It appears to be much less frequent than that species in Somerset and to be found on more acid soils. It has definitely been collected in districts 2, 3, 9 and 10.

Carex echinata Murr. Star Sedge

Wet moorland and marshy places on acid soils. Common on the western moors, the Blackdown Hills, the eastern borders, peat moors and the Mendip sandstone. Rare elsewhere. Not recorded in district 5.

Carex remota L. Remote Sedge

By streams, on ditchsides and in damp woods. Very common throughout the county.

†**Carex curta** Gooden. White Sedge

This plant of acid bogs, which is now very rare in the south of England, was recorded on the peat moors (8) by Sole (1791) but has not been seen there since.

Carex ovalis Gooden. Oval Sedge

Wet moorland, woodland rides and wet meadows. Common on acid soils. Rather rare elsewhere but recorded in all districts.

Carex pulicaris L. Flea Sedge

In boggy places. Frequent on moorland and other acid soils. Rather rare and decreasing due to drainage elsewhere. Not recorded in district 5.

†**Carex dioica** L. Dioecious Sedge

Spongy bogs. Collected by Jenyns on Shapwick Moor (8) in 1855 but not seen there since. It must always have been very rare there as it is not mentioned by any earlier botanist. There is also a record by Mr Walker "near Bath" in Babington (1834), said by T. B. Flower to have been in North Somerset and within four miles of the city but nothing more is known of this.

†**Carex davalliana** Sm.

The only British locality known for this Central European species was a bog on Lansdown on the outskirts of Bath (10) where it was found by a Mr Groult. Specimens from this locality are in the herbaria at the British Museum (ante 1809) and Cambridge University (1813). The place was drained (and has since been built over) and Babington wrote in 1834 that it had not been seen "for some years".

GRAMINEAE

Leersia Sw.

Leersia oryzoides (L.) Sw. Cut-grass

Discovered in the Bridgwater & Taunton Canal near North Newton (3) in 1959 by my wife and myself. Clumps were found at intervals along a stretch of some four miles of the canal. It still grows in many of the spots where it was first seen but perhaps not in such quantity as in 1959 which was a very good year for it; many panicles were fully exserted and it flowered and fruited freely which does not happen every year.

Phragmites Adans.

Phragmites australis (Cav.) Trin. ex Steud. Common Reed

P. communis Trin.

In ditches, in and by rivers and ponds. Common throughout the Levels and frequent elsewhere on the lower ground. Not recorded in district 1.

Molinia Schrank

Molinia caerulea (L.) Moench Purple Moor-grass

Wet heaths and moors and sometimes in wet open woodland. Common on acid and peaty soils. Rare on calcareous ones and only where locally acid conditions have developed on clay. Not recorded recently in district 7.

Danthonia DC.

Danthonia decumbens (L.) DC. Heath Grass

Sieglingia decumbens (L.) Bernh.

On moors, heaths and poor hill grassland. Frequent in such habitats and recorded in all districts.

Glyceria R. Br.

Glyceria fluitans (L.) R. Br. Floating Sweet-grass

In ditches, by streams and in other watery places. Very common throughout the county.

× **plicata** (*G.* × *pedicillata* Townsend). A sterile hybrid which sometimes occurs in the absence of both parents and which has been recorded in all districts except 1 and 6.

Glyceria plicata Fr. Plicate Sweet-grass

In ditches, ponds and slow streams. Very common in the eastern half of the county. Rather rare in the west and not recorded recently in district 1.

Glyceria declinata Bréb. Small Sweet-grass

By pools and in muddy places where water has stood. More frequent in the west and on acid soils than elsewhere but widespread and recorded in all districts. Not distinguished by the earlier botanists but Marshall studied it for many years; most of his records are from the western half of the county.

Glyceria maxima (Hartm.) Holmberg Reed Sweet-grass

G. aquatica (L.) Wahlb.

In ditches, by canals, slow rivers and ponds. Very common throughout the Levels, choking drainage ditches unless kept under control. Rather rare elsewhere especially in the west.

FESTUCA L.

Festuca pratensis Huds. Meadow Fescue

F. elatior var. *pratensis* sensu Murray

In meadows and on grassy roadsides. Common throughout the county.

× **Lolium perenne** (× *Festulolium loliaceum* (Huds.) P. Fourn.). Frequent in rich meadows in most parts of the county.

Festuca arundinacea Schreb. Tall Fescue

F. elatior L. nom. ambig.

Roadsides, rough grassland and waste places. Common except on Exmoor where there are few records.

× **Lolium perenne** (× *Festulolium holmbergii* (Dörfl.) P. Fourn.). A rather rare or overlooked hybrid which has been recorded in both V.C.5, at Sparkford (5), 1967 (CAH), and in V.C.6, in the Avon Gorge (10), 1955 (C & NS). In both cases the records were confirmed by C. E. Hubbard.

Festuca gigantea (L.) Vill. Giant Fescue

Bromus giganteus L.

Woods and shady hedgebanks. Very common except on the barer parts of Exmoor.

Festuca nigrescens Lam. Chewings Fescue

F. rubra ssp. *commutata* Gaud.; *F. fallax* Thuill.

An important constituent of grass seed sown on lawns and roadsides. Very common but its native distribution in the county is uncertain.

Festuca rubra L. Red Fescue

F. duriuscula auct.; *F. oraria* sensu White

Grassland, cliffs, sand dunes, salt marshes, walls and waste places. Very common everywhere. A very variable species now divided into a number of subspecies five of which are believed to occur in Somerset but their respective distributions have not yet been worked out.

× **Vulpia fasciculata** (× *Festulpia hubbardii* Stace & Cotton). This hybrid was not identified in Britain until 1954 and in 1957 it was collected on Berrow dunes (9) by P. C. & J. F. Hall. Since then it has appeared in other places in Britain where *Vulpia fasciculata* grows and this has been attributed to the reduction in rabbits from myxomatosis increasing the opportunities for hybridization to occur.

Festuca ovina L. Sheep's Fescue

On moors, downs and in hilly grassland. Very common in all districts.

LOLIUM L.

Lolium perenne L. Perennial Rye-grass

In meadows, on roadsides and in waste ground. Very common everywhere and an important constituent of sown grass mixtures.

Lolium multiflorum Lam. Italian Rye-grass

L. italicum A. Braun

An introduced species much sown for hay or grazing and now common on roadsides and waste ground in all parts of the county except the extreme west. A branched form (forma **ramosa** (Guss.) P. Junge) is sometimes found.

× **perenne** (*L.* × *hybridum* Hausskn.). The two species interbreed freely and hybrids are said to be common but have only been identified positively in Somerset on waste ground and rubbish tips in the Bristol area (10).

Lolium temulentum L. Darnel

Formerly a weed of arable land but now only a fairly frequent casual of waste ground and rubbish tips.

Lolium rigidum Gaud.

A Mediterranean grass found on a tip at Brislington (10), 1978 (ALG).

VULPIA C. C. Gmel.

Vulpia fasciculata (Forsk.) Samp. Dune Fescue

V. membranacea auct.; *Festuca uniglumis* Ait.; *Lolium bromoides* Huds.

Sand dunes. Rare and only at Burnham (8) and Berrow (9). Recorded also near Minehead (2) by W. H. Coleman in 1849 but not seen there since.

Vulpia bromoides (L.) Gray Squirrel-tail Fescue

Festuca sciuroides Roth

On dry banks and wall tops, in poor grassland and waste places. Frequent throughout the county.

Vulpia myuros (L.) C. C. Gmel. Rat's-tail Fescue

Festuca myuros L.

Roadsides and dry waste places. Frequent on railway ground. Rare elsewhere but recorded in all districts except 1.

Vulpia ciliata Dumort. ssp. **ambigua** (Le Gall) Stace & Auquier Bearded Fescue

V. ambigua (Le Gall) More

Sand dunes. Very rare and only at Berrow (9) where it was first found by Mrs C. I. and N. Y. Sandwith in 1960.

†Vulpia unilateralis (L.) Stace — Mat-grass Fescue

Nardurus maritimus (L.) Murb.

This rare plant of thin grassland on limestone was found in a clearing in a wood on the Lias near Pitminster (3) in 1949 by Miss M. McC. Webster but it has since disappeared, apparently shaded out by woodland growth. There was also a casual occurrence on a rubbish tip at Ashton Gate (10) in 1937 (C & NS).

PUCCINELLIA Parl.

Puccinellia maritima (Huds.) Parl. — Common Saltmarsh Grass

Glyceria maritima (Huds.) Wahlb.; *Sclerochloa maritima* (Huds.) Lindl. ex Bab.

Saltmarshes. Common all along the coast from Porlock (2) to the Avon estuary below Bristol (10).

Puccinellia distans (L.) Parl. — Reflexed Saltmarsh Grass

Glyceria distans (L.) Wahlenb.; *Sclerochloa distans* (L.) Bab.

On mud and in waste places near the sea and river estuaries and occasionally on rubbish tips further inland. Rather rare near the coast between Minehead (2) and Portbury (10) and on tips near Glastonbury (8), 1976 (CAH), and Brislington (10), 1979 (ALG).

Puccinellia rupestris (With.) Fernald & Weatherby — Stiff Saltmarsh Grass

Festuca procumbens Kunth.; *Sclerochloa procumbens* Beauv.

In muddy and stony places near the sea. Very rare and decreasing. It has been seen recently only at Kewstoke, Portishead (9), Portbury and Pill (10). It also grows sparsely on the north bank of the New Cut at Bristol (10), (V.C.34). Formerly also at Bridgwater (3 & 5) and Clevedon (9) but not reported from either for many years.

DESMAZERIA Dumort.

Desmazeria rigida (L.) Tutin — Fern Grass

Catapodium rigidum (L.) C. E. Hubbard; *Festuca rigida* (L.) Rasp.; *Sclerochloa rigida* (L.) Link

On and under walls and in other dry places. Common over most of the county except on Exmoor where it is rare.

GRAMINEAE

Desmazeria marina (L.) Druce — Sea Fern Grass

Catapodium marinum (L.) C. E. Hubbard; *Sclerochloa loliacea* Woods; *Festuca loliacea* auct.

On rocks, walls and in sandy places near the sea. Fairly frequent between Brean Down and Sand Point (9). Very rare elsewhere the only recent records being at Bossington and Minehead (2), Combwich (3), Huntspill (8) and on Steep Holm. There are old records from Stolford, Steart (2) and Berrow (9).

Poa L.

Poa annua L. — Annual Meadow Grass

Grassland, cultivated ground and waste places. A very common weed everywhere.

Poa bulbosa L. — Bulbous Meadow Grass

Sandy places near the sea. Very rare. Found for the first time in the county in 1979 on the beach at Dunster (2) by Miss C. J. Giddens (conf. CEH). Old reports from Weston-super-Mare (9) and Newton St Loe (10) are believed to have been errors. A patch of the var. **vivipara** Koch was found in an old quarry on Hampton Down (10) in 1975 (R & IR); this may have been derived from grass seed sown on the adjoining golf course.

Poa nemoralis L. — Wood Meadow Grass

Woods and shady banks. Frequent and recorded in all districts, but rather scarce in the central lowlands. Introduced in many ornamental woodlands and plantations and apparently more widespread than in the past; Murray (1896) classed it as "rare and confined to the north of the county" which is certainly not so now.

Poa compressa L. — Flattened Meadow Grass

On walls and in other dry places. Widespread and locally common particularly in the eastern half of the county. Recorded in all districts.

Poa pratensis L. — Smooth Meadow Grass

Grassland, roadsides, waste places and wall tops. Very common throughout the county.

Poa angustifolia L. Narrow-leaved Meadow Grass

P. pratensis ssp. *angustifolia* (L.) Gaudin

Rough grassland, roadsides and railway banks on calcareous soils. Frequent on the Lias and Oolite. Rare elsewhere and not recorded in districts 1 and 6, nor recently in 3.

Poa subcaerulea Sm. Spreading Meadow Grass

P. pratensis ssp. *subcaerulea* (Sm.) Tutin

Damp meadows and dune slacks. Widespread but not common. The distribution of this species in the county has not been fully worked out because of identification difficulties.

Poa trivialis L. Rough Meadow Grass

In meadows, woods, cultivated land, waste places and on roadsides. Very common in all districts.

Catabrosa Beauv.

Catabrosa aquatica (L.) Beauv. Whorl Grass

Muddy margins of pools and ditches and in shallow water. Rare and apparently decreasing but perhaps sometimes overlooked as it is eagerly grazed by cattle if they can reach it. Recent records are scattered about the county in districts 2, 3, 5, 8, 9 and 10 but there are older ones in districts 1 and 4 also.

Dactylis L.

Dactylis glomerata L. Cocksfoot

In grassland, on roadsides and in waste places. Very common everywhere.

Cynosurus L.

Cynosurus cristatus L. Crested Dog's-tail

In grassland. Very common throughout the county.

Cynosurus echinatus L. Rough Dog's-tail

A Mediterranean grass which was a frequent casual about the Docks some years ago but which has only once been seen recently, at Lower Swainswick, Bath (10), 1976 (TC).

GRAMINEAE

Briza L.

Briza media L. Quaking Grass

In old grassland. Common on dry calcareous soils. Less so on more acid ones and then usually in marshy places. Recorded in all districts.

†**Briza minor** L. Lesser Quaking Grass

Although stated to occur near Bath (10) as long ago as 1778 a localized record given by Babington (1834) was later admitted by him to have been a mistake. Included for North Somerset in *Topographical Botany* (1873-74) on the authority of Withers but it is unlikely the plant ever grew in the county except as a casual.

Briza maxima L. Great Quaking Grass

Garden escape.

Melica L.

Melica uniflora Retz. Wood Melick

M. nutans auct.

Woods and shady hedgebanks. Common except on the Levels where there are few suitable habitats.

Bromus L.

Bromus erectus Huds. Upright Brome

Zerna erecta (Huds.) Gray

Calcareous grassland. Common on the downs of North Somerset from the Polden Hills northwards. Very rare in the west and south. There are no records in district 1 and very few in 2, 3, 4 and 6.

Bromus ramosus Huds. Hairy Brome

Zerna ramosa (Huds.) Lindm.

Woods and shady hedgebanks. Common throughout the county except on the higher parts of Exmoor.

Bromus sterilis L. Barren Brome

Anisantha sterilis (L.) Nevski

Roadsides, hedgebanks, walls and in waste places. Very common in all districts except on the highest parts of the hills.

Bromus madritensis L. Compact Brome

Anisantha madritensis (L.) Nevski

On limestone rocks and dry waste ground. Very rare and as a native only in the Avon Gorge (10). It was reported on Stert Island (2) in 1960 (JVM) and has persisted there but a few other reports, in districts 4, 9 and 10, have proved ephemeral.

Bromus diandrus Roth Great Brome

Anisantha gussonii (Parl.) Nevski

An introduced grass which has become established on sandy soils in some parts of the country. It has been reported twice in Somerset, at Dunster (2), 1977 (JA & RGR), and at Berrow (9), 1961 (NYS), but it has not persisted in either place. Both identifications were confirmed by C. E. Hubbard.

Bromus tectorum L. Drooping Brome

Anisantha tectorum (L.) Nevski

A rare casual, last recorded on a wall at Weston-in-Gordano (9) in 1966 (RMH). It had been reported only twice previously in the county.

Bromus hordeaceus L. Soft Brome

B. mollis L.

Meadows, roadsides and waste ground. Very common as an aggregate throughout the county. Ssp. **thominii** (Hardouin) Hylander, a coastal plant with rather dense small panicles, has been identified on Brean Down and at Uphill (9), 1970 (PJMN det. CEH). Ssp. **ferronii** (Mabile) P. Smith, a cliff-top grass with stiffly erect peduncles and dense erect panicles, has also been reported from Brean Down, 1968 (T. C. E. Wells conf. A. Melderis) but could not be found in 1970 (PJMN).

× **lepidus** (*B.* × *pseudothominii* P. Smith; *B. thominii* auct.). In sown grassland, hayfields, pastures and on roadsides. The hybrid origin of this plant has only recently been recognized. It is believed it is introduced in grass-seed mixtures. Frequent and widespread in the county.

Bromus lepidus Holmberg Slender Soft Brome

B. britannicus I. A. Williams

This introduced grass was not generally recognized as distinct from *B.*

hordeaceus until 1929. The earliest known Somerset specimens were collected on Tickenham Hill (9) by Miss I. M. Roper in 1915. It is a not infrequent casual of grassland and roadsides and like its hybrid with *B. hordeaceus* is introduced in grass-seed mixtures, the seeds being much the same size as those of the Rye-grasses.

Bromus racemosus L. Smooth Brome

Serrafalcus racemosus (L.) Parl.

In old meadows and on roadsides. The distribution is rather uncertain because of confusion with *B. commutatus;* it is apparently quite frequent in the eastern half of the county but it has not been recorded recently in the western half.

Bromus commutatus Schrad. Meadow Brome

Serrafalcus commutatus (Schrad.) Bab.

In meadows and on roadsides. Common on the Levels especially on the grassy droves. Rare elsewhere and not seen recently in districts 1 or 2.

Bromus secalinus L. Rye Brome

Serrafalcus secalinus (L.) Bab.

Cornfields, roadsides and waste places. An introduced species formerly a frequent weed of cornfields and on rubbish tips but now rare. It has been recorded recently in cornfields on the Lias in districts 4, 5 and 8.

Bromus willdenowii Kunth Rescue Brome

B. unioloides auct. non Kunth; *Ceratochloa unioloides* (Willd.) Beauv.

A fodder plant from America, first recorded in the county in 1897 at Twerton (10), which became quite frequent about the Docks by 1932 (CIS) but then disappeared again. Only once recorded recently, on a tip at Brislington (10), 1979 (ALG).

Bromus lanceolatus Roth

B. macrostachys Desf.

A wool alien which appeared in quantity in gardens and waste places at New Town, Yeovil (4) in 1977 (JGK det. CEH).

BRACHYPODIUM Beauv.

Brachypodium sylvaticum (Huds.) Beauv. False Brome

Woods and hedgebanks. Very common in all districts except on the higher parts of Exmoor.

Brachypodium pinnatum (L.) Beauv. Tor Grass

In dry grassland on limestone hills. Rare and local. On the Polden Hills near Butleigh (5), on Mendip from Loxton Hill to Burrington (9), at Cadbury Camp, Tickenham (9) and in several places on the hills round Bath (10).

ELYMUS L.

Elymus caninus (L.) L. Bearded Couch

Agropyron caninum (L.) Beauv.; *Triticum caninum* L.

Wood borders, shady hedgebanks and by rivers. Frequent in the south and east and common in the north of the county. Not recorded in the west in districts 1 or 2. It has longer and more flexuous awns than *E. repens* var. *aristatum* which is sometimes mistaken for it.

Elymus repens (L.) Gould Common Couch

Agropyron repens (L.) Beauv.; *Triticum repens* L.

In arable land, waste places and on hedgebanks. Very common throughout the county. Awned varieties are frequent; the length of the awns varies and may be up to 10 mm. long.

Elymus pycnanthus (Godr.) Melderis Sea Couch

Agropyron pycnanthum (Godr.) Gren. & Godr.; *A. pungens* auct.; *Triticum pungens* auct.

Salt marshes and the muddy banks of tidal rivers. Common from Porlock (2) to the Avon estuary (10).

Elymus farctus (Viv.) Runemark ex Melderis Sand Couch
Ssp **boreali-atlanticus** (Simonet & Guinochet) Melderis

Agropyron junceiforme (A. & D. Löve) A. & D. Löve; *A. junceum* auct.; *Triticum junceum* auct.

On sandy shores. Locally common at Minehead and Steart (2) and between Burnham (8) and Kewstoke (9).

× **pycnanthus** (*Agropyron* × *obtusiusculum* Lange; *A. acutum* auct.) has been reported in the past from places where *E. farctus* grows but not identified recently at any of them. It was also found near Rownham Ferry (10) by Miss Attwood (Swete, 1854) but was thought to have gone from the Avon estuary until 1976 when it was rediscovered below Leigh Woods (Mrs O. M. Stewart conf. CEH).

LEYMUS Hochst.

Leymus arenarius (L.) Hochst. — Lyme Grass

Elymus arenarius L.

On sand dunes. Local and rare. Recently recorded only at Minehead (2), Stert Island (2), Burnham (8) and Berrow (9). Formerly also at Weston-super-Mare and Kewstoke (9) but not seen in either place this century.

SECALE L.

Secale cereale L. — Rye

An occasional casual or relic of cultivation.

TRITICUM L.

Triticum aestivum L. — Wheat

Commonly cultivated and a frequent casual by roadsides, in farmyards, etc.

Triticum durum Desv. — Macaroni Wheat

A casual, first reported in the county in 1978 on a tip at Brislington (10) (ALG & TGE det. EJC).

HORDEUM L.

Hordeum secalinum Schreb. — Meadow Barley

H. nodosum auct.

In damp meadows on clay. Very common throughout the Levels and frequent elsewhere on the lower ground. Not recorded in districts 1 or 6.

Hordeum murinum L. Wall Barley

A weed of dry waste places especially near the sea and in the larger towns. Common in most parts of the county except on the higher ground. Not recorded in district 1. Its rarity about Bath (10) was remarked on by Sole (1799) and many other botanists including White (1912) but it is now quite common there.

Hordeum marinum Huds. Sea Barley

H. maritimum Stokes

In salt marshes and on sea banks. Frequent. It occurs at Blue Anchor (2) and although both Murray (1896) and White (1912) describe it as "rather rare" it is common enough now all along the coast between Hinkley Point (2) and Weston-super-Mare (9) and from Wick St Lawrence to Clevedon (9). It is also at Portbury (9) and Easton-in-Gordano (10).

Hordeum hystrix Roth

This, with **H. murinum** ssp. **glaucum** (Steud.) Tzvelev, **H. pubiflorum** Hook. f. and other wool adventives, were found in 1970 (CAH, CES & JGK) on a tip near Glastonbury (8) used by the local leather factory. The factory no longer uses foreign skins so these species are not likely to recur.

Hordeum jubatum L.

A North American species occasionally found on rubbish tips.

Hordeum vulgare L. Barley

Hordeum distichon L.

Both these species of Barley are commonly cultivated and are frequently found as casuals on roadsides.

KOELERIA Pers.

Koeleria macrantha (Ledeb.) Schultes Crested Hair-grass

K. cristata (L.) Pers. pro parte; *K. gracilis* Pers.; *K. britannica* Domin; *Aira cristata* L.

Calcareous grassland. Rare in South Somerset and only recorded recently at Old Cleeve (2), Ham Hill (4) and South Cadbury (5). Frequent in North Somerset and quite common on the Polden Hills (5), Mendips (9) and the hills round Bath (10).

× **vallesiana.** This hybrid was first detected in 1933 by W. O. Howarth at Uphill and on Crook Peak (9). A survey in 1974 by R. S. Callow established that it exists in extremely small numbers, with both parent species, on Brean Down, Worle Hill and Sand Point (9) also. Plants collected at Uphill in 1905 by G. C. Druce and named as *K. mixta* Domin are almost certainly variants of *K. macrantha* and not this hybrid.

Koeleria vallesiana (Honck.) Gaud. Somerset Hair-grass

On south-facing rocky limestone slopes. Confined in Britain to a small area of the North Somerset Carboniferous Limestone near the sea but locally abundant there. It grows from Brean Down at intervals eastwards along the southern slope of Mendip as far as Fry's Hill, Axbridge, a distance of some ten miles, and also on Worle Hill and Sand Point (all in district 9). This grass was first collected by Dillenius at Brean Down and Uphill on 16th July, 1726, but his memorandum and specimens became separated and remained unnoticed in the Sherardian herbarium until in October, 1904, G. C. Druce came on them while working on a memoir of Dillenius. He quickly verified that the plant still grew at Uphill. In 1967 C. E. Hubbard found, when going through the Rev J. Lightfoot's herbarium at Kew, a specimen labelled "*Aira cristata,* Weston supra Mare" which was probably collected on his visit to Brean Down on 3rd July, 1773. The overlooking of this species for so long by the many botanists who have visited Brean Down over the centuries is surprising; even after its rediscovery it was not until 1958 that it was first reported east of Crook Peak (E. J. Hope Simpson) and 1974 before it was found on Sand Point (R. S. Callow).

Gaudinia Beauv.

Gaudinia fragilis (L.) Beauv.

This Mediterranean species was first recorded in Britain in 1903 and in Somerset as a casual in 1928 at Bristol and Bedminster (10) (C & NS). In 1970 it was found by J. G. Keylock near Haselbury Plucknett (4) along a ride in a Forestry Commission plantation; in 1972 he found it in a new ley about a mile away and in 1974 in a permanent pasture some two miles from where it was first seen. It appears to be thoroughly established in the last locality which is probably the source of the earlier sightings.

TRISETUM Pers.

Trisetum flavescens (L.) Beauv. Yellow Oat-grass

T. pratense Pers.

In dry grassland and on roadsides. Very common throughout the county except in the extreme west.

AVENA L.

Avena fatua L. Wild Oat

A weed of cornfields and arable land. Common in most districts and increasing. The reduction in other weeds of arable land due to the use of weedkillers has been accompanied by an increase in this pest. Not recorded yet in district 1.

Avena sterilis L. ssp. **ludoviciana** (Durieu) Nyman Winter Wild Oat

A. ludoviciana Durieu

This plant, which has become a serious pest of cereal crops on heavier soils in parts of the Midlands since its introduction into Britain during the 1914-1918 War, is still only a rare casual in Somerset. It was recorded near Bath (10) in 1962 and 1963 (RGR), near Stawell (5), 1977 (RGR) and in several spots near Charlton Mackrell (5) from 1976 onwards. It may well be more frequent but overlooked as it is difficult to tell from *A. fatua* until the grain is ripe.

Avena sativa L. Oat

A frequent relic of cultivation.

Avena barbata Pott. ex Link

A wool alien once recorded in a sheepskin-storage yard at Glastonbury (8), 1971 (CAH det. CEH).

AVENULA (Dumort.) Dumort.

Avenula pratensis (L.) Dumort. Meadow Oat-grass

Helictotrichon pratense (L.) Pilg.; *Avena pratensis* L.

Grassland on calcareous downs. Frequent on the limestones in the eastern half of the county but rare in the west and there only on the Lias. Not recorded in district 1.

GRAMINEAE

Avenula pubescens (Huds.) Dumort. Downy Oat-grass

Helictotrichon pubescens (Huds.) Pilg.; *Avena pubescens* Huds.

Calcareous grassland. Its distribution is similar to that of *A. pratensis* but it is more common and occurs in meadows on damper and deeper soils as well as on downland. Not recorded recently in district 1.

ARRHENATHERUM Beauv.

Arrhenatherum elatius (L.) Beauv. ex J. & C. Presl False Oat

A. avenaceum Beauv.

Rough grassland, hedgebanks, roadsides and waste places. Very common in all districts.

HOLCUS L.

Holcus lanatus L. Yorkshire Fog

Meadows and pastures, roadsides and waste places. Very common everywhere in all types of grassland.

× **mollis** (*H.* × *hybridus* K. Wein). Apparently a rare hybrid in Britain but possibly overlooked. Records by W. Watson made in 1918 from several localities were not accepted by Marshall and no confirmed record has been made in the county since.

Holcus mollis L. Creeping Soft-grass

Open woodland, heaths, hedgebanks and a weed in poor, sandy soil. Common on acid soils. Rather rare elsewhere and absent from much of the Levels.

DESCHAMPSIA Beauv.

Deschampsia cespitosa (L.) Beauv. Tufted Hair-grass

Aira cespitosa L.

Damp woods, marshy grassland and by ditches. Very common throughout the county.

Deschampsia flexuosa (L.) Trin. Wavy Hair-grass

Moors, heaths and dry open woodland on acid soils. Common on the moorland and hills of the west, on the eastern Greensand and on the Mendip sandstone. Very rare elsewhere. Not recorded recently in district 5.

AIRA L.

Aira praecox L. Early Hair-grass

In dry sandy or gravelly places on acid soils. Common in the west. Rather rare elsewhere but recorded in all districts.

Aira caryophyllea L. Silver Hair-grass

In similar places to *A. praecox* but more widespread and not so restricted to acid soils. Not recorded recently in district 7.

AMMOPHILA Host

Ammophila arenaria (L.) Link Marram

A. arundinacea Host; *Arundo arenaria* L.

On sand dunes, Minehead Warren, Steart (2) and from Burnham (8) to Brean, Weston-super-Mare and Kewstoke (9). Often planted to consolidate drifting sand but it was recorded between Burnham and Brean by Sole (1791). Its arrival at Minehead may be more recent as there are no records from there before 1930.

CALAMAGROSTIS Adans.

Calamagrostis epigejos (L.) Roth Wood Small-reed

In damp woods and by roadsides on calcareous clay. Frequent in the eastern half of the county. Rare except on the Lias in the west. Not recorded in districts 1 or 6.

(**Calamagrostis canescens** (Weber) Roth is credited to North Somerset in Watson's *Topographical Botany* on the strength of records by Sole and Davis quoted by Babington (1834). These records almost certainly referred to *C. epigejos.* Sole's original record in Collinson (1791) is for *"Arundo calamagrostis"* which was the name used by Hudson and other British botanists for *C. epigejos* until 1804 when Smith correctly applied it to *C. canescens.* Sole died in 1802 and cannot be blamed for the error.)

AGROSTIS L.

Agrostis curtisii Kerguélen Bristle Bent

A. setacea Curt.

On moors and heaths. Locally common on Exmoor and the Quantocks (1, 2 & 3), Haddon Hill (1), Langford Heathfield (3) and the Blackdown Hills (3, 4 & 6).

GRAMINEAE

Agrostis canina L. Brown Bent

Heaths, moors and dry grassland. Common on acid soils only, so absent from many parts of the county but recorded in all districts.

Agrostis capillaris L. Common Bent

A. tenuis Sibth.; *A. vulgaris* With.

Grassland and waste places. Very common in all districts especially on acid soils.

Agrostis gigantea Roth Black Bent

A. nigra With.

A weed of arable land, waste places and roadsides. Formerly rather rare but it has become common and a serious pest in some places. Not yet recorded in district 1.

Agrostis stolonifera L. Creeping Bent

A. alba auct.

Grassland. Very common on all types of soil throughout the county.

× **Polypogon monspeliensis** (× *Agropogon littoralis* (Sm.) C. E. Hubbard; *Polypogon littoralis* Sm.). A rare hybrid found with the parent species and other salt-marsh grasses on a disused tip near Glastonbury (8) in 1973 (CAH det. CEH). Although sterile the grass is a perennial and was still there in 1979.

Agrostis avenacea J. F. Gmel.

A wool alien found on a tip at Glastonbury (8) in 1970 (CAH & CES) while it was still in use for dumping waste from a leather factory.

POLYPOGON Desf.

Polypogon monspeliensis (L.) Desf. Annual Beard-grass

A salt-marsh grass only recorded as a casual in Somerset. Recent records are at Wellington (3), Glastonbury (8), Portbury (9) and Brislington (10).

GASTRIDIUM Beauv.

Gastridium ventricosum (Gouan) Schinz & Thell. Nit Grass

G. australe Beauv.

On shallow, dry, unstable soil on limestone slopes. Very rare and only recorded recently at Old Cleeve (2), 1968 (C. A. Stace) and Aller (5), 1975 (H. J. M. Bowen). Old records for "Minehead" may refer to the former locality. There are other old records from near Wells (8) and Bath (10) but it has not been seen near either for many years. Reports from Corfe (3) and Purn Hill, Bleadon (9) were probably of casual occurrences.

LAGURUS L.

Lagurus ovatus L. Hare's-tail

A rare casual first recorded in the county near Berrow (9), 1929 (WDM), and only seen recently on tips at Weston Bampfylde (5), 1969 (CAH), and Brislington (10), 1979 (ALG).

PHLEUM L.

Phleum pratense L. ssp. **pratense** Timothy Grass

Meadows and pastures. Very common throughout the county. Native only in the lowlands but commonly sown.

Ssp. **bertolonii** (DC.) Bornm. Smaller Cat's-tail

P. bertolonii DC.; *P. nodosum* auct.

Grassland. Common in old pastures, particularly on downs, throughout the county.

Phleum arenarium L. Sand Cat's-tail

On dunes and in other sandy places near the sea. Minehead (2) and along the coast from Burnham (8) to Kewstoke (9). Formerly recorded also from Steart (2) and Clevedon (9). A very old record from Tickenham (9) has never since been confirmed.

ALOPECURUS L.

Alopecurus myosuroides Huds. Black Twitch

A weed of cultivated land and roadsides. Common on the calcareous clays of central Somerset and frequent elsewhere. Not recorded in districts 1 or 6.

GRAMINEAE

Alopecurus pratensis L. Meadow Foxtail

Meadows and pastures. Very common over most of the county except on the highest ground.

Alopecurus geniculatus L. Marsh Foxtail

In wet meadows, marshes and by pools and ditches. Common in the marshlands and frequent elsewhere.

× **pratensis** (*A.* × *brachystylus* Peterm.; *A.* × *hybridus* Wimm.). A rare or overlooked hybrid reported from Norton Fitzwarren (3) in 1922 (WDM).

(**Alopecurus aequalis** Sobol. (*A. fulvus* Sm.) was included for Somerset with no authority named in *Topographical Botany* (1873-74) but never since confirmed.)

Alopecurus bulbosus Gouan Bulbous Foxtail

Salt marshes and brackish meadows near the sea. Although declining elsewhere this rare grass is locally plentiful in Somerset and has apparently increased. Murray (1896) gives only one locality, near Dunster (2), and White (1912) two more, near Portishead (9) and below Pill (10). It is still at Blue Anchor (2) and by the Avon below Pill (10) and it has since been found at Stolford (2), 1946 (AD & OMH), Wall Common (2), 1968 (HWB), Combwich (3), 1967 (HWB), Pawlett (5), 1974 (RSC), Highbridge (8), 1923 (H. J. Gibbons) and Lympsham (9), 1976 (RSC).

MILIUM L.

Milium effusum L. Wood Millet

In woods and plantations. Frequent. Sometimes sown as food for game birds. Not recorded recently in district 6.

PIPTATHERUM Beauv.

Piptatherum miliaceum (L.) Cosson

Oryzopsis miliacea (L.) Benth. & Hook. ex Adans. & Schweinf.

A rare casual or escape first recorded in the county on a tip at Brislington (10), 1978 (ALG).

ANTHOXANTHUM L.

Anthoxanthum odoratum L. Sweet Vernal Grass

Meadows, pastures, woods and heaths. Very common throughout the county especially on neutral and acid soils.

PHALARIS L.

Phalaris arundinacea L. Reed Canary-grass

By rivers, ditches and ponds. Common in all districts. The var. **picta** L., with cream-striped leaves is an occasional garden escape.

Phalaris canariensis L. Canary-grass

A frequent casual on rubbish tips and in waste places.

Phalaris paradoxa L.

A casual once frequent in the Bristol area but only seen recently on a tip at Brislington (10), 1978 (ALG & TGE conf. CEH).

PARAPHOLIS C. E. Hubbard

Parapholis strigosa (Dumort.) C. E. Hubbard Hard-grass

Lepturus filiformis auct.; *L. incurvatus* auct.

Salt marshes by the sea and tidal estuaries. Porlock, Bossington, Lilstock and locally common from Hinkley Point (2) to Berrow and Uphill (9), from Wick St Lawrence to Clevedon and from Portishead (9) to the Avon estuary at Pill (10). It occurs up the Parrett as far as Bridgwater (5). Formerly also in other spots along the coast and up the Avon as far as Bristol (10).

Parapholis incurva (L.) C. E. Hubbard Curved Hard-grass

Pholiurus incurvus (L.) Schinz & Thell.

Salt marshes. Very rare. Not correctly distinguished by British botanists until fairly recently, this grass was first detected in Somerset in 1938 at Berrow (R. Melville) and Brean (9) (P. J. M. Brenan). Possibly overlooked because of difficulty in separating it from *P. strigosa,* the only recent record in the county was at Bossington (2), 1964 (P. Rogerson).

NARDUS L.

Nardus stricta L. Mat-grass

Moors and heaths on acid soils. Common on Exmoor (1 & 2) and the Quantocks (2 & 3). Less so on the Brendons (1), Langford Heathfield (3), the Blackdown Hills (3, 4 & 6), the eastern Greensand (7 & 10) and the Mendip sandstone (8 & 9). Not recorded recently on the peat moors (8) nor on the sandstone west of Bristol (9 & 10).

GRAMINEAE

SPARTINA Schreb.

Spartina alterniflora Lois. × **S. maritima** (Curt.) Fernald
Townsend's Cord-grass

S. × *townsendii* H. & J. Groves

The fertile amphidiploid derived by chromosome doubling from the F_1 hybrid was first collected at Lymington, Hampshire, in 1892 and soon became established in tidal mud-flats on the south coast. About a thousand tufts were brought from Poole Harbour in 1913 and planted on the foreshore between the mouth of the River Yeo and Clevedon (9). In the next four years the colony doubled in size. In 1921 plants were recorded at Berrow (9) by W. Watson and H. S. Thompson independently. In 1929 more was planted in Bridgwater Bay (2) and by 1933 it had spread to Burnham (8) and Sand Bay (9). It was first recorded at Portishead (9) in 1946 and at Uphill (9) in 1951. The date of its arrival in the Avon estuary (10) is not known but it now extends upwards as far as the Gorge. It is still very scarce in the west; it was reported at Lilstock (2) in 1966 (JVM) and at Porlock Weir (2) in 1974 (CJG), but from Bridgwater Bay (2) to the Avon estuary (10) it is now abundant on the mud-flats. The sterile F_1 hybrid has been identified in Sand Bay (9) and in the Avon estuary and probably occurs here and there among the amphidiploid plants. The latter are now called **S. anglica** C. E. Hubbard to distinguish them.

CYNODON Rich.

Cynodon dactylon (L.) Pers. Bermuda Grass

An introduced grass which can become established on sandy ground and in waste places. It was known on waste ground at Bath (10) for fifty years but disappeared about 1946. A small patch was found on Minehead Warren (2) in 1973 by Miss C. J. Giddens and this has since spread to cover quite a large area. The grass was noticed on a railway bank at Weston-super-Mare (9) in 1976 by R. H. Walters and in 1979 R. M. Payne reported that it was extensively naturalized on the lawns on the sea front there. It is likely to persist in both these places. It has also occurred as a casual at Burnham (8) and Glastonbury (8).

ECHINOCHLOA Beauv.

Echinochloa crus-galli (L.) Beauv. Cockspur

A frequent casual on rubbish tips and in waste places.

Echinochloa colonum (L.) Link

A tropical weed, first recorded as a casual in Somerset in 1978 on a rubbish tip at Brislington (10) (ALG, CML & AT).

Echinochloa frumentacea Link — Indian Millet

Echinochloa utilis Ohwi & Yabano — Japanese Millet

These Asiatic cereals were first recorded in the county on a tip near Shepton Mallet (8) in 1973 (CAH & CES det. CEH) and are becoming increasingly frequent on similar dumps of town rubbish.

DIGITARIA Haller

Digitaria sanguinalis (L.) Scop. — Hairy Finger-grass

A casual reported in gardens and waste places from time to time since 1896 but only seen twice recently, at Brislington (10), 1978 (ALG & TGE), and Bath (10), 1978 (RDR).

Digitaria ciliaris (Retz.) Koel.

A rare casual first found in the county on a tip at Brislington (10) in 1978 (ALG, CML & AT det. CEH).

SETARIA Beauv.

Setaria viridis (L.) Beauv. — Green Bristle-grass

A frequent casual of tips and waste ground, known in the county since 1881.

Setaria verticillata (L.) Beauv. — Rough Bristle-grass

A rare casual, first recorded in the county at Burnham (8), 1927 (WDM), and only seen recently on a tip at Brislington (10), 1978 (ALG, CML & AT).

Setaria pumila (Poir.) Schult. — Yellow Bristle-grass

S. glauca auct.; *S. lutescens* (Weig.) F. T. Hubbard

Formerly a casual of gardens and waste ground but now only found on rubbish tips, probably derived from bird seed, as is also **Setaria italica** (L.) Beauv. which is more common.

GRAMINEAE

Setaria geniculata (Lam.) Beauv.

A rare casual once found on a tip at Brislington (10), 1978 (ALG & TGE det. EJC).

PANICUM L.

Panicum capillare L.

Casual, first recorded in the county at St Catherine, near Bath (10), 1955 (G. W. Garlick det. CEH) and since seen only on a tip at Brislington (10), 1978 (ALG, CML & AT conf. EJC).

Panicum miliaceum L. Millet Grass

A bird-seed alien common on rubbish tips.

Panicum dichotomiflorum Michx

A rare casual found on the Brislington tip (10), 1979 (ALG & TGE det. CEH).

SORGHUM Moench

Sorghum halepense (L.) Pers.

An oriental casual seen at Clevedon (9), (G. W. Garlick det. CEH), and on a tip at Brislington (10), (ALG, CML & AT), in 1978.

ZEA L.

Zea mays L. Maize

Increasingly cultivated in Somerset as cattle fodder and sometimes found on tips and in waste places.

DIPLACHNE Beauv.

Diplachne uninervia (Presl.) Parodi

A rare casual found on the Brislington tip (10), 1978 (ALG, CML & AT).

ERAGROSTIS N. M. Wolf

Eragrostis cilianensis (All.) F. T. Hubbard

Eragrostis neomexicana Vasey

Rare casuals found on the Brislington tip (10), 1978 (ALG, CML & AT).

APPENDIX

EXTINCT CASUALS

The following casuals and escapes have been recorded in the past but have not been seen in Somerset since 1960. Most were found about docks, railway yards, mills and warehouses, and many are unlikely to recur because of the changes in trade and commercial practices. The dates are those of the last known record.

ASPIDIACEAE

Dryopteris villarii (Bellardi) Woynar (*Lastraea rigida* (Sw.) C. Presl) 1866

RANUNCULACEAE

Consolida orientalis (Gay) Schrödinger (*Delphinium orientale* J. Gay) 1932
Consolida regalis S. F. Gray (*Delphinium consolida* L.) 1932
Clematis flammula L. 1936
Ranunculus muricatus L. 1897
Ranunculus fluitans Lam. 1917
Ranunculus trilobus Desf. 1915
Adonis annua L. 1938
Adonis aestivalis L. 1907

PAPAVERACEAE

Argemone mexicana L. 1956
Roemeria hybrida (L.) DC. 1956
Glaucium corniculatum (L.) Rudolph (*G. phoeniceum* Crantz) 1938
Glaucium grandiflorum Boiss. & Huet 1936
Hypecoum procumbens L. 1907
Hypecoum pendulum L. 1907
Corydalis solida (L.) Sw. (*Capnoides solida* (L.) Moench) 1910
Corydalis bulbosa (L.) DC. (*Capnoides cava* (L.) Schwiegg. & Koerte) 1922

CRUCIFERAE

Brassica elongata Ehrh. 1929
Brassica tournefortii Gouan 1949
Brassica pollichii Sch. & Sp. 1916
Diplotaxis erucoides (L.) DC. 1919

APPENDIX

Carrichtera annua (L.) DC. (*C. vellae* DC.)	ante 1896
Raphanus sativus L.	1932
Rapistrum perenne (L.) All.	1923
Conringia orientalis (L.) Dumort.	1932
Lepidium virginicum L.	1938
Lepidium bonariense L.	1922
Lepidium densiflorum Schrad.	1927
Lepidium neglectum Thell.	1940
Lepidium perfoliatum L.	1932
Myagrum perfoliatum L.	1922
Iberis odorata L. (*I. acutiloba* Bertol.)	1922
Neslia paniculata (L.) Desv. (*Vogelia paniculata* (L.) Hornem.)	1932
Bunias erucago L.	1940
Lunaria rediviva L.	1932
Alyssum alyssoides (L.) L.	1898
Alyssum minus (L.) Rothm. (*A. campestre* auct.)	1907
Alyssum hirsutum Bieb.	1897
Berteroa incana (L.) DC.	1915
Draba aizoides L.	ante 1912
Cardamine bellidifolia L.	ante 1805
Euclidium syriacum (L.) R. Br.	1940
Barbarea stricta Andrz.	1903
Matthiola incana (L.) R. Br.	ante 1896
Matthiola longipetala (Vent.) DC ssp. bicornis (Sibth. & Sm.) P. W. Ball (*M. bicornis* (Sibth. & Sm.) DC.)	1935
Malcolmia africana (L.) R. Br.	1914
Erysimum virgatum Roth	ante 1842
Erysimum repandum L.	1928
Sisymbrium polyceratium L.	1922
Sisymbrium multifidum (Pursh.) McMill.	1922
Camelina microcarpa Andrz. ex DC. (*C. sylvestris* Wallr.)	1922
Camelina alyssum (Mill.) Thell.	1937
Camelina sagittata Crantz	1938
RESEDACEAE	
Reseda inodora Reichb.	1936
TAMARICACEAE	
Tamarix tetrandra Pallas ex Bieb.	1918
CARYOPHYLLACEAE	
Silene conica L. ssp. subconica (Friv.) Gavioli (*S. subconica* Friv.)	1939

APPENDIX

Silene conoidea L.	1906
Silene dichotoma Ehrh.	1932
Silene armeria L.	1920
Silene nutans L.	1834
Silene cretica L.	1915
Silene muscipula L.	1932
Silene stricta L.	1922
Silene behen L.	1941
Lychnis preslii Sekera	1929
Gypsophila acutifolia Stev. ex Spreng.	1952
Gypsophila elegans Bieb.	1922
Gypsophila pilosa Huds. (*G. porrigens* (L.) Boiss.)	1907
Dianthus caryophyllus L.	ante 1829
Dianthus arenarius L.	ante 1791
Petrorhagia velutina (Guss.) P. W. Ball & Heywood (*Dianthus velutinus* Guss.)	1902
Corrigiola litoralis L.	1896
Herniaria hirsuta L.	1914

AMARANTACEAE

Amaranthus caudatus L.	1940
Amaranthus spinosus L.	ante 1960
Amaranthus graecizans L. (*A. sylvestris* Vill.)	ante 1960
Amaranthus deflexus L.	1957
Amaranthus lividus L. (*A. blitum* L.)	1911
Amaranthus viridis L. (*A. gracilis* Desf.)	1932
Achyranthes aspera L.	ante 1960

CHENOPODIACEAE

Chenopodium multifidum L. (*Roubieva multifida* (L.) Moq.)	1918
Chenopodium foliosum Aschers. (*Blitum virgatum* L.)	1918
Chenopodium capitatum (L.) Aschers.	1930
Chenopodium opulifolium Schrad. ex Koch & Ziz	1909
Chenopodium hircinum Schrad.	1928
Chenopodium pratericola Rydb. (*C. leptophyllum* auct.)	1921
Chenopodium procerum Hochst. ex Moq.	ante 1960
Monolepis nuttalliana (Schult.) Greene	1917
Atriplex rosea L.	1922
Atriplex tatarica L.	1917
Axyris amaranthoides L.	1936
Salsola kali L. ssp. ruthenica (Iljin) Soó (*S. pestifer* A. Nels.)	1930

APPENDIX

Malvaceae	
Malva nicaeensis All.	ante 1960
Malva pusilla Sm. (*M. borealis* Wallr.)	1957
Malva verticillata L.	ante 1960
Lavatera punctata All.	1922
Anoda cristata (L.) Schlecht.	1939
Abutilon theophrasti Medic. (*A. avicennae* Gaertn.)	1919
Hibiscus trionum L.	ante 1960
Geraniaceae	
Geranium sylvaticum L.	ante 1914
Geranium nodosum L.	1929
Geranium reflexum L.	1929
Zygophyllaceae	
Tribulus terrestris L.	1931
Oxalidaceae	
Oxalis valdiviensis Barnéoud	1959
Celastraceae	
Euonymus latifolius (L.) Mill.	1918
Leguminosae	
Lupinus angustifolius L.	1897
Medicago scutellata (L.) Mill.	1897
Medicago truncatula Gaertn. (*M. tribuloides* Desr.)	1897
Medicago aculeata Gaertn. (*M. turbinata* Willd.)	1907
Medicago laciniata (L.) Mill.	1897
Trigonella polyceratia L.	1922
Trigonella monspeliaca L.	1922
Trigonella procumbens (Besser) Reichb. (*T. besserana* Ser.)	1932
Trigonella foenum-graecum L.	1921
Trigonella coelesyriaca L.	1930
Trifolium ochroleucon Huds.	1897
Trifolium aureum Poll.	1924
Trifolium nigrescens Viv.	1916
Trifolium angulatum Waldst. & Kit.	1918
Trifolium spumosum L.	1907
Trifolium vesiculosum Savi	1914
Trifolium diffusum Ehrh.	1922
Trifolium cherleri L.	1897
Trifolium angustifolium L.	1918
Trifolium alexandrinum L.	1897

APPENDIX

Colutea arborescens L.	1951
Astragalus stella Gouan	1930
Ornithopus compressus L.	1897
Vicia villosa Roth ssp. varia (Host) Corb. (*V. dasycarpa* Ten.)	1917
Vicia sativa L. ssp. macrocarpa (Moris) Arcangeli (*V. macrocarpa* Bert.)	1916
Vicia melanops Sibth. & Sm.	1922
Lathyrus inconspicuus L.	1940
Lathyrus cicera L.	1922
Lathyrus sativus L.	1913
Lathyrus annuus L.	1913
Lathyrus ochrus (L.) DC.	1922
ROSACEAE	
Fragaria moschata Duchesne	ante 1912
Pyracantha coccinea M. J. Roem.	1959
Pyrus pyraster Burgsd.	1936
CRASSULACEAE	
Sedum sexangulare L.	1917
Sedum stellatum L.	1912
GROSSULARIACEAE	
Ribes spicatum Robson	1920
Ribes alpinum L.	1891
PUNICACEAE	
Punica granatum L.	1949
ONAGRACEAE	
Oenothera laciniata Hill (*O. sinuata* L.)	1922
Oenothera rosea L'Hér. ex Ait.	1918
UMBELLIFERAE	
Scandix stellata Banks & Sol. (*S. pinnatifida* Vent.)	1921
Scandix iberica Bieb.	1917
Torilis leptophylla (L.) Reichan b.f. (*Caucalis leptophylla* L.)	1922
Bifora testiculata (L.) Roth	1940
Bifora radians Bieb.	1940
Apium leptophyllum (Pers.) F. Muell. ex Benth.	1923
Falcaria vulgaris Bernh.	ante 1896
Levisticum officinale Koch	ante 1896
EUPHORBIACEAE	
Euphorbia dulcis L.	1947

APPENDIX

Euphorbia cyparissias L. 1921
Euphorbia characias L. 1953

POLYGONACEAE
Polygonum arenarium Waldst. & Kit. ssp. pulchellum (Lois.) D. A. Webb & Chater (*P. pulchellum* Lois.) 1930
Polygonum corrigioloides Janb. & Spach. 1926
Fagopyrum tataricum (L.) Gaertn. 1942
Rumex scutatus L. 1958
Rumex patientia L. 1942
Rumex dentatus L. 1928

URTICACEAE
Girardinia condensata (Hochst. ex Steud.) Wedd. ante 1960

PLUMBAGINACEAE
Limonium suworowi (Regel) O. Kuntze 1942

PRIMULACEAE
Androsace maxima L. 1909

OLEACEAE
Ligustrum lucidum Aiton f. (*L. japonicum* auct.) ante 1912

POLEMONIACEAE
Polemonium caeruleum L. 1918
Collonia linearis Nutt. 1922
Gilia capitata Sims 1922
Plagiobothrys canescens Benth. 1917

BORAGINACEAE
Lappula squarrosa (Retz.) Dumort. (*Echinospermum lappula* (L.) Lehm. 1926
Amsinckia lycopsoides (Lehm.) Lehm. ante 1912
Amsinckia calycina (Moris) Chater (*A. angustifolia* Lehm.) 1921
Amsinckia intermedia Fisch. & C. A. Meyer 1932
Symphytum tuberosum L. 1956
Anchusa officinalis L. 1917
Anchusa undulata L. ssp. hybrida (Ten.) Coutinho (*A. hybrida* Ten.) 1940
Anchusa stylosa Bieb. 1939
Echium italicum L. 1917

CONVOLVULACEAE
Ipomoea purpurea Roth 1928
Cuscuta approximata Bab. 1906

APPENDIX

Solanaceae	
Physalis ixocarpa Brot. ex Hornem.	ante 1960
Physalis foetens Poir.	1926
Solanum triflorum Nutt.	1917
Nicotiana rustica L.	1912
Scrophulariaceae	
Verbascum densiflorum Bertol. (*V. thapsiforme* Schrad.)	1919
Verbascum pulverulentum Vill.	1959
Verbascum chaixii Vill.	1912
Verbascum ovalifolium Donn ex Sims (*V. pulchrum* Velen.)	1932
Verbascum phoeniceum L.	1911
Linaria pelisseriana (L.) Mill.	1886
Chaenorhinum origanifolium (L.) Fourr. (*Linaria origanifolia* (L.) Cav.)	ca. 1900
Scrophularia umbrosa Dumort.	1897
Parentucellia viscosa (L.) Caruel	1950
Orobanchaceae	
Orobanche ramosa L.	1778
Verbenaceae	
Verbena tenera Spreng.	1932
Labiatae	
Salvia sclarea L.	1958
Salvia glutinosa L.	1938
Salvia pratensis L.	1919
Salvia nemorosa L. (*S. sylvestris* auct.)	1932
Stachys annua (L.) L.	1908
Stachys recta L.	1898
Stachys byzantina C. Koch (*S. lanata* Jacq.)	1912
Lamium moluccellifolium Fr.	1907
Wiedemannia orientalis Fisch. & Meyer	1916
Marrubium peregrinum L.	1923
Sideritis montana L.	1932
Plantaginaceae	
Plantago sempervirens Crantz (*P. cynops* L.)	1926
Plantago aristata Michx	1922
Plantago lagopus L.	1907
Campanulaceae	
Campanula rapunculus L.	ante 1896

APPENDIX

Legousia speculum-veneris (L.) Chaix (*Specularia speculum-veneris* (L.) A. DC.)	1909
Lobelia dortmanna L.	1918
CAPRIFOLIACEAE	
Lonicera japonica Thunb.	1952
DIPSACACEAE	
Cephalaria syriaca (L.) Roem. & Schult.	1897
Scabiosa atropurpurea L.	1873
COMPOSITAE	
Rudbeckia laciniata L.	1880
Lepachys columnaris Torr. & Gray	1918
Bidens pilosa L.	ante 1960
Bidens bipinnata L.	ante 1960
Helianthus rigidus × H. tuberosus (*H.* × *laetiflorus* Pers.)	1935
Ambrosia coronopifolia Torr. & Gray (*A. psilostachya* auct.)	1917
Ambrosia trifida L.	1938
Xanthium chinense Mill.	ante 1960
Hemizonia pungens (Hook. & Am.) Torr. & Gray	1921
Hemizonia kelloggii Greene	1922
Senecio inaequidens DC.	ante 1960
Cacalia hastata L.	1906
Encelia mexicana Mart.	1900
Calendula arvensis L.	1918
Asteriscus aquaticus (L.) Less.	1897
Erigeron annuus (L.) Pers. ssp. strigosus (Muhl. ex Willd.) Wagenitz (*E. strigosus* Muhl. ex Willd.)	1942
Anthemis ruthenica Bieb.	1923
Anthemis wiedemanniana Fisch. &Meyer	1939
Chamaemelum mixtum (L.) All.	1923
Anacyclus clavatus (Desf.) Pers.	1926
Anacyclus valentinus L.	1926
Chrysanthemum coronarium L.	1929
Tanacetum corymbosum (L.) Schultz Bip. (*Chrysanthemum corymbosum* L.)	1928
Artemisia gnaphalodes Nutt.	1927
Mantisalca salmantica (L.) Briq. & Cav. (*Centaurea salmantica* L.)	1922
Centaurea iberica Trev. ex Spreng. ssp. holzmanniana (Boiss.) Dostál	1958
Centaurea solstitialis L.	1938
Centaurea melitensis L.	1932

APPENDIX

Cnicus benedictus L.	1897
Carthamus lanatus L.	1937
Madia sativa Molina	1948
Madia glomerata Hook.	1923
Hedypnois cretica (L.) Dum.-Courset (*H. rhagadioloides* (L.) F. W. Schmidt)	1939
Scorzonera hispanica L.	1958
Lactuca saligna L.	1868
Hieracium caespitosum Dumort. ssp. colliniforme (Peter) P. D. Sell (*H. colliniforme* (Naeg. & Peter) Roffey; *H. pratense* auct.)	1917
Crepis foetida L.	1939
Crepis nicaeensis Balbis	1883
Crepis zacintha (L.) Babcock (*Zacintha verrucosa* Gaertn.)	1913
Hydrocharitaceae	
Lagarosiphon major (Ridl.) Moss	1959
Liliaceae	
Asphodelus fistulosus L. (*A. tenuifolius* Cav.)	1932
Ornithogalum nutans L.	1926
Allium schoenoprasum L.	1957
Allium moly L.	1957
Iridaceae	
Hermodactylus tuberosus (L.) Mill. (*Iris tuberosa* L.)	1950
Crocus vernus (L.) Hill (*C. purpureus* Weston)	1918
Palmae	
Phoenix dactylifera L.	1932
Gramineae	
Festuca altissima All. (*F. sylvatica* (Pollich) Vill.)	ante 1886
Festuca heterophylla Lam.	1959
Festuca lemanii Bast. (*F. longifolia* auct.)	1959
Poa palustris L.	1948
Poa chaixii Vill.	1959
Beckmannia eruciformis (L.) Host	1926
Beckmannia syzigachne (Steud.) Fernald	1930
Bromus inermis Leyss. (*Zerna inermis* (Leyss.) Lindm.)	1926
Bromus brachystachys Hornung	1897
Bromus interruptus (Hack.) Druce	1915
Bromus arvensis L. (*Serrafalcus arvensis* (L.) Godr.)	1922
Bromus japonicus Thunb. (*Serrafalcus patulus* Parl.)	1931
Bromus scoparius L.	1939

APPENDIX

Bromus squarrosus L.	1897
Aegilops speltoides Tausch (*A. bicornis* auct.)	1926
Aegilops cylindrica Host	1921
Aegilops triuncialis L.	1939
Aegilops ventricosa Tausch	1907
Aegilops neglecta Req. ex Bertol. (*A. triaristata* Willd.)	1906
Aegilops ligustica (Sav.) Coss.	1938
Elymus canadensis L.	1929
Taeniatherum caput-medusae (L.) Nevski (*Elymus caput-medusae* L.)	1930
Hordeum trifurcatum	1923
Koeleria berythaea Boiss. & Blanche	1939
Lophochloa cristata (L.) Hyl. (*Koeleria phleoides* (Vill.) Pers.)	1939
Avena strigosa Schreb.	1930
Zingeria pisidica (Boiss.) Tutin (*Agrostis pisidica* Boiss.)	1940
Apera spica-venti (L.) Beauv.	1931
Apera interrupta (L.) Beauv.	1931
Phleum hirsutum Honck. (*P. michelii* All.)	1907
Phalaris minor Retz	ante 1960
Phalaris angusta Nees	1928
Phalaris brachystachys Link	1930
Phalaris coerulescens Desf. (*P. aquatica* auct.)	1930
Phalaris cylindracea DC.	1907
Chloris pycnothrix Trin.	ante 1960
Chloris ventricosa R. Br.	1915
Panicum effusum R. Br.	1941
Panicum laevifolium Hack.	ante 1960
Brachiaria isachne (Roth) Stapf	1935
Paspalum racemosum Lam.	1922
Stipa hyalina Nees	1928
Snowdenia polystachya (Fresen.) Pilger	ante 1960
Dactyloctenium radulans (R. Br.) Beauv.	1915
Eleusine africana O'Byrne	ante 1960
Eleusine floccifolia (Forsk.) Spreng.	ante 1960
Eragrostis pilosa (L.) Beauv.	1926
Eragrostis tef (Zucc.) Trotter	ante 1960

BIBLIOGRAPHY AND REFERENCES

BABINGTON, C. C. (1834). *Flora Bathoniensis.*

BABINGTON, C. C. (1839). *Supplement to Flora Bathoniensis.*

BAKER, J. G. (1875). On the rarer plants of Central Somersetshire. *J. Bot.* **13,** 357-361.

BENTHAM, G. & HOOKER, J. D. (1904). *Handbook of the British Flora,* edn. 8.

Botanical Society and Exchange Club of the British Isles. *Reports.*

Botanist's Guide. See Turner & Dillwyn (1805).

Bristol Naturalists' Society, Proceedings. (1917-). Bristol Botany.

B.S.B.I. Atlas. See Perring & Walters (1962).

CLAPHAM, A. R., TUTIN, T. G. & WARBURG, E. F. (1962). *Flora of the British Isles.* edn. 2.

CLARK, T. (1858). Catalogue of the rarer plants of the Turf Moors of Somerset. *Proc. Som. Arch. Soc.* **7(2),** 64-71.

COLEMAN, W. H. (1849). List of plants seen within ten miles of Minehead and Dunster supplied to Watson (1873-74).

COLLINS, J. C. (1837). List of plants supplied to Watson (1837).

COLLINSON, J. (1791). *The History of Somersetshire.*

DANDY, J. E. (1958). *List of British Vascular Plants.*

DAVIS, J. F. (1802). List of plants in Warner (1802).

DONY, J. G., PERRING, F. & ROB, C. M. (1974). *English names of Wild Flowers.*

DRUCE, G. C. (1932). *The Comital Flora of the British Isles.*

DUCK, J. N. (1852). *Natural History of Portishead.*

DUNN, S. T. (1897). Aliens from Turkish Barley. *J. Bot.* **35,** 444.

EDWARDS, Z. J. (1865). *The Ferns of the Axe and its tributaries.*

GAPPER, A. (1835). List of plants supplied to Watson (1835).

GERARD, J. (1597). *The Herball.*

GERARD, J. (1633). *The Herball,* edn. 2.

GIDDENS, C. J. (1979). *Flowers of Exmoor,* revised edition.

GIFFORD, I. (1856). Notices of the rare and remarkable plants in the neighbourhood of Blue Anchor, Minehead, etc. *Proc. Som. Arch. Soc.* **6(2),** 131-137.

GROSE, D. (1957). *The Flora of Wiltshire.*

HUDSON, W. (1778). *Flora Anglica,* edn. 2.

JENYNS, L. (1867). On the Bath Flora. *Proc. Bath Nat. Hist. & Antiq. Field Club* **1,** 23-48.

JOHNSON, T. (1634). *Mercurius Botanicus.*

Journal of Botany (J. Bot.)

LOUSLEY, J. E. (1961). A census list of wool aliens found in Britain 1946-60. *Proc. bot. soc. Br. Isl.* **4,** 221-147.

MARSHALL, E. S. (1907-19). Notes on Somerset Plants. *J. Bot.* **45-57.**

MARSHALL, E. S. (1914). *A Supplement to the Flora of Somerset.*

MARTIN, W. K. & FRAZER, G. T., eds. (1939). *Flora of Devon.*

BIBLIOGRAPHY AND REFERENCES

MILLER, W. D. (1933). A note on extinct and rare species of the county of Somerset. *Rep. botl Soc. Exch. Club Br. Isl.* **10,** 268-276.

MURRAY, R. P. (1896). *The Flora of Somerset.*

New Botanist's Guide. See Watson (1835-37).

NEWMAN, E. (1854). *History of British Ferns,* edn. 3.

PARSONS, H. F. (1876). The Flora of the eastern border of Somersetshire. *Proc. Som. Arch. Soc.* **21(2),** 53-61.

PERRING, F. H. & WALTERS, S. M. eds. (1962). *Atlas of the British Flora.*

PERRING, F. H. ed. (1968). *Critical Supplement to the Atlas of the British Flora.*

POLWHELE, R. (1797). *History of Devonshire.*

RAY, J. (1670). *Catalogus Plantarum Angliae.*

RAY, J. (1688). *Historia Plantarum,* Vol. 2.

RAY, J. (1690). *Synopsis Methodica Stirpium Britannicarum,* edn. 1.

RAY, J. (1724). *Synopsis Methodica Stirpium Britannicarum,* edn. 3 (by Dillenius).

RICHARDS, A. J. (1972). The *Taraxacum* flora of the British Isles. Supplement to *Watsonia,* **9.**

RUTTER, J. (1829). *Delineation of the north-western division of the county of Somerset.*

ST BRODY, G. (1856). *Flora of Weston.*

SANDWITH, C. I. (1933). Adventive flora of the Port of Bristol. *Rep. botl Soc. Exch. Club Br. Isl.* **10,** 314-363.

SOLE, W. (1782). Ms Flora of Somerset (unpublished).

SOLE, W. (1791). List of "the more rare plants growing in this county" in Collinson (1791).

SOLE, W. (1801). List of plants in Warner (1801).

Somersetshire Archaeological and Natural History Society, Proceedings. (1910-) Botanical notes and records.

STACE, C. A. ed. (1975). *Hybridization and the flora of the British Isles.*

STEARN, L. F. ed. (1975). *Supplement to the Flora of Wiltshire.*

STEPHENS, H. O. (1835). Catalogue of plants in the neighbourhood of Bristol. *West of England Journal.*

SWETE, E. H. (1854). *Flora Bristoliensis.*

SYME, J. T. B. (1863-86). *English Botany,* edn. 3.

Topographical Botany. See Watson (1873-74).

TURNER, D. & DILLWYN, L. W. (1805). *The Botanist's Guide through England and Wales.*

TURNER, W. (1562). *A newe Herball,* part 2.

TURNER, W. (1568). *A newe Herball,* part 3.

TUTIN, T. G. et al. eds. (1964-80). *Flora Europaea.* Vol. 1 (1964). Vol. 2 (1968). Vol 3 (1972). Vol. 4 (1976). Vol. 5 (1980).

WARNER, R. (1801). *History of Bath.*

WARNER, R. (1802). *An historical and descriptive account of Bath and its environs.*

WATSON, H. C. (1835). *The New Botanist's Guide to the localities of the rarer plants of Britain.*

WATSON, H. C. (1837). Vol. 2 and Supplement.

WATSON, H. C. (1873-74). *Topographical Botany.*

BIBLIOGRAPHY AND REFERENCES

WHITE, J. W. (1886). *Flora of the Bristol Coal-field.*
WHITE, J. W. (1912). *The Flora of Bristol.*
WHITE, J. W. (1918). Notes supplemental to the Flora of Bristol. *J. Bot.* **56,** 11-
WITHERING, W. (1796). *Botanical Arrangement of British Plants,* edn. 3.

INDEX

INDEX

INDEX

INDEX

INDEX

INDEX

INDEX

INDEX

INDEX

INDEX

INDEX

INDEX

INDEX

INDEX

INDEX

INDEX

INDEX

INDEX

INDEX

INDEX

INDEX

INDEX

INDEX

INDEX

INDEX

INDEX

INDEX

INDEX

INDEX

NOTES

NOTES

NOTES

NOTES

NOTES